AF595034

State-of-the-Art Research in Biomolecular Crystals

State-of-the-Art Research in Biomolecular Crystals

Editor

Abel Moreno

MDPI • Basel • Beijing • Wuhan • Barcelona • Belgrade • Manchester • Tokyo • Cluj • Tianjin

Editor
Abel Moreno
Universidad Nacional
Autónoma de México
Mexico

Editorial Office
MDPI
St. Alban-Anlage 66
4052 Basel, Switzerland

This is a reprint of articles from the Special Issue published online in the open access journal *Crystals* (ISSN 2073-4352) (available at: https://www.mdpi.com/journal/crystals/special_issues/biomolecular_crystals).

For citation purposes, cite each article independently as indicated on the article page online and as indicated below:

LastName, A.A.; LastName, B.B.; LastName, C.C. Article Title. *Journal Name* **Year**, *Volume Number*, Page Range.

ISBN 978-3-0365-6454-8 (Hbk)
ISBN 978-3-0365-6455-5 (PDF)

Cover image courtesy of Abel Moreno

Contents

About the Editor

Abel Moreno

Dr. Abel Moreno was awarded a B.Sc. in Chemistry from the Autonomous University of Puebla (Mexico) in 1990 and Ph.D. in Chemistry from the University of Granada (Spain) in 1995. Currently, Dr. Moreno is a Full Professor of Biological and Physical Chemistry at the Institute of Chemistry of the National Autonomous University of Mexico (UNAM) in Mexico City. He has been distinguished as a member of the National System of Researchers of Mexico (SNI) at level 3 (the highest category of Mexican scientists) and a member of the Mexican Academy of Sciences, Mexican Society of Crystallography, Mexican Society of Synchrotron Light, the New York Academy of Sciences, and the Mexican and American Chemical Societies as well as the Spanish Royal Society of Chemistry. He has been a visiting professor at the University of Cambridge (United Kingdom, January–December 2009) and at the University of Strasbourg (France, 2003–2004).

He is the author of 15 book chapters and 7 books on his specialties in biological crystallogenesis, crystallochemistry, and biomineralization processes. He is a member of the Editorial Board of the journal *Progress in Crystal Growth and Characterization of Materials* (ELSEVIER), Editor for the Latin America section of the Newsletter of the International Union of Crystallography, and Editor-in-Chief of the section Biomolecular Crystals of the journal *Crystals* (MDPI, Switzerland). He was recently appointed as Associate Editor of the journal *CrystEngComm* from the RSC (UK) as well as the *Mexican Journal of Physics* (Mexico).

Editorial

State-of-the Art Research in Biomolecular Crystals

Abel Moreno

Instituto de Química, Universidad Nacional Autónoma de México, Avenida Universidad 3000, Colonia UNAM, Ciudad de Mexico 04510, Mexico; carcamo@unam.mx; Tel.: +52-55-56224467

This special issue, State-of-the Art Investigations on Biomolecular Crystals, is focused on strategies to procure suitable crystals for high-resolution X-ray crystallographic investigations [1,2] from experimental methods related to the use of different biomolecules for applications in biomedical sciences or bioinorganic chemistry [3]. The structure of these biomolecules, performed by X-ray diffraction, even using synchrotron radiation, combines different applications of peptides, proteins, and DNAs crystals in medicine, biology and in materials science. These techniques, employing the use of synchrotron facilities, will mold a new kind of future protein crystallographers trained in using novel software with big data recorded. Additionally, the tools for manipulating micro or nanocrystals are completely new and will also be included in this special issue.

X-rays, electron, and neutron diffraction techniques have been the most ubiquitously used strategies to obtain the structural elucidation of biomolecules and biomacromolecular complexes. Recently, with the development of synchrotron facilities, we necessitate smaller sizes of crystals progressing from microns to picometers. Exposing biological cryo-protected crystals to the X-ray intensity in a shorter time, combined with the use of highly efficient photon detectors, diminishes radiation damage. However, the use of small angle X-ray scattering techniques (SAXS) has developed into an extremely popular method for obtaining the structural envelope in solution for any biological macromolecule, where the structure is arduous to be solved due to the necessity of crystals [1,3]. Nuclear magnetic resonance (NMR) was also extremely popular in the 90s (of the previous century) for obtaining protein structures in solution, although with a limitation in the molecular weight (MW). This limitation weight occurred because the larger the MW, the lower the possibility to procure the structure at very high resolution (the one-dimensional or two-dimensional NMR spectra are much more multifactorial with regards interpretation) [4]. X-ray crystallography is nowadays not only the principal method, but it also the most puissant method, for obtaining three-dimensional structures of macromolecules at a very high resolution, independently of their molecular weight. Particularly, in the classical crystallographic approach, X-ray diffraction necessitates high-quality single crystals for high resolution. We must take into consideration that possessing beautiful crystals does not necessarily signify that we will procure high-resolution structures in merely a few steps. Occasionally, ugly-shaped crystals, which are internally well structured, produce the appropriate X-ray patterns to solve the three-dimensional structure at excellent resolution. Currently, the existence of other methods to obtain the three-dimensional structure of any biological macromolecule from the investigations on nucleation up to the crystal growth have make it plausible to labor in the direct space, such as the case with cryo-electron microscopy (cryo-EM) or reciprocal space when using the classic X-ray crystallographic approaches [5]. The other possibility will be the use of linear synchrotrons of the fourth generation such as the XFEL, where the data collection does not strictly necessitate any type of high-quality or large crystals but, rather, highly pure proteins, nucleic acids or macromolecular complexes in solution are needed. We can employ even a few microliters of a viscous dispersion of micro and nanocrystals for injection bypassing through the pulses of high energy to obtain the three-dimensional structure of any biomolecule before it is destroyed [6,7]. Besides these techniques for obtaining the three-dimensional structure of a variety of

Citation: Moreno, A. State-of-the Art Research in Biomolecular Crystals. *Crystals* **2023**, *13*, 58. https://doi.org/10.3390/cryst13010058

Received: 16 December 2022
Revised: 16 December 2022
Accepted: 23 December 2022
Published: 29 December 2022

proteins, nucleic acids or polysaccharides, as well as their complexes, there is a want of physicochemical mechanisms in crystal formation and protein–protein interactions (based on pH, temperature, or ionic strength). The significance of the crystal quality [8–10], the presence of impurities [11], the insights into the crystallization process from protein nucleation phenomena, crystallization, and crystal growth peculiarities are important topics for review [12]. The existence of polymorphs is another research topic that deserves to be investigated in detail for crystals of biological macromolecules [13]. The use of specific crystallization agents or organic polymers, such as poly-ethylene glycols (400–20,000 MW) and their interaction with proteins, will be discussed in one of the contributions associated with this special issue. The measurements of the crystal growth rate, either on Earth or in microgravity, have recently shown interesting information in terms of crystal quality and mechanisms of crystal growth [14].

The main intention of the the state-of-the art in biomolecular crystals was to present a collection of papers where diverse applications of crystals were investigated. The first group is dedicated to biomedical applications for Parkinson's disease [15,16]; the second group is concerned with the application of crystals for basic research, combining organic (DNA) and inorganic crystals (silica carbonate biomorphs of Ba^{2+}) [17]; theoretical models using PEGs for protein crystallization [18] and the importance of polymorphs are also considered [19]. The third group deals with crystal fracture breakage [20], cross-linked enzyme crystals [21], stabilization of DNA-Protein co-crystals [22], biotechnological approaches using the amidase process [23] and binding of covalent inhibitors, and reversible ligands and substrates [24].

We hope that this collection of papers for this special issue will inspire a plethora of publications regarding these trending topics soon.

Funding: The author is thankful to DGAPA-UNAM for the financial support (grant number IN207922), which, to some extent, has facilitated him towards being guest editor of this Special Issue.

Acknowledgments: The author acknowledges the support from DGAPA-UNAM, project IN207922, for the support regarding the crystallization and structural investigations on human transferrins mentioned in this work and for using SAXS methods. Additionally, the author thanks MDPI editor for their help and professional support concerning the design and completion of this special issue.

Conflicts of Interest: The funders had no role in the design of the study; in the collection, analyses, or interpretation of data; in the writing of the manuscript; or in the decision to publish the results.

References

1. Campos-Escamilla, C.; González-Ramírez, L.A.; Otálora, F.; Gavira, J.A.; Moreno, A. A short overview on practical techniques for protein crystallization and a new approach using low intensity electromagnetic fields. *Prog. Cryst. Growth Charact. Mater.* **2022**, *68*, 100559. [CrossRef]
2. Gavira, J.A. Current trends in protein crystallization. *Arch. Biochem. Biophys.* **2016**, *602*, 3–11. [CrossRef] [PubMed]
3. Campos-Escamilla, C.; Siliqi, D.; González-Ramírez, L.A.; López-Sánchez, C.; Gavira, J.A.; Moreno, A. X-ray Characterization of Conformational Changes of Human Apo- and Holo-Transferrin. *Int. J. Mol. Sci.* **2021**, *22*, 13392. [CrossRef] [PubMed]
4. Cavalli, A.; Salvatella, X.; Dobson, C.M.; Vendruscolo, M. Protein Structure Determination from NMR Chemical Shifts. *Proc. Natl. Acad. Sci. USA* **2007**, *104*, 9615–9620. [CrossRef] [PubMed]
5. Benjin, X.; Ling, L. Developments, applications, and prospects of cryo-electron microscopy. *Protein Sci.* **2020**, *29*, 872–882. [CrossRef] [PubMed]
6. Boutet, S.; Lomb, L.; Williams, G.J.; Barends, T.R.M.; Aquila, A.; Doak, R.B.; Weierstall, U.; DePonte, D.P.; Steinbrener, J.; Shoeman, R.L.; et al. High-resolution protein structure determination by serial femtosecond crystallography. *Science* **2012**, *337*, 362–364. [CrossRef]
7. Fromme, R.; Ishchenko, A.; Metz, M.; Chowdhury, S.R.; Basu, S.; Boutet, S.; Fromme, P.; White, T.A.; Barty, A.; Spence, J.C.H.; et al. Serial femtosecond crystallography of soluble proteins in lipidic cubic phase. *IUCrJ* **2015**, *2*, 545–551. [CrossRef]
8. Lee, K.M.; Bae, S.H.; Park, J.I.; Kwon, S.O. Synchrotron X-ray reciprocal-space mapping, topography and diffraction resolution studies of macromolecular crystal quality. *Acta. Crystallogr. D Biol. Crystallogr.* **2000**, *56*, 868–880.
9. Otalora, F.F.; Garcia-Ruiz, J.M.; Gavira, J.A.; Capelle, B. Topography and high-resolution diffraction studies in tetragonal lysozyme. *J. Cryst. Growth* **1999**, *196*, 546–558. [CrossRef]
10. Robert, M.-C.; Capelle, B.; Lorber, B. Growth Sectors and Crystal Quality. *Methods Enzymol.* **2003**, *368*, 154–169.

11. Robert, M.-C.; Capelle, B.; Lorber, B.; Giegé, R. Influence of impurities on protein crystal perfection. *J. Cryst. Growth* **2001**, *232*, 489–497. [CrossRef]
12. Nanev, C.N. Recent Insights into Protein Crystal Nucleation. *Crystals* **2018**, *8*, 219. [CrossRef]
13. Gillespie, C.M.; Asthagiri, D.; Lenhoff, A.M. Polymorphic Protein Crystal Growth: Influence of Hydration and Ions in Glucose Isomerase. *Cryst. Growth Des.* **2014**, *14*, 4657. [CrossRef] [PubMed]
14. Tsukamoto, K.; Furukawa, E.; Dold, P.; Yamamoto, M.; Tachibana, M.; Kojina, K.; Yoshizaki, I.; Vlieg, E.; González-Ramírez, L.A.; Gracía-Ruiz, J.M. Higher Growth Rate of Protein Crystals in Space than on Earth. *J. Cryst. Growth* **2023**, *603*, 127016. [CrossRef]
15. Chochkova, M.; Rusew, R.; Kalfin, R.; Tancheva, L.; Lazarova, M.; Sbirkova-Dimitrova, H.; Popatanasov, A.; Tasheva, K.; Shivachev, B.; Petek, N.; et al. Synthesis, Molecular Docking, and Neuroprotective Effect of 2-Methylcinnamic Acid Amide in a 1-methyl-4-phenyl-1,2,3,6-tetrahydropyridine (MPTP)-An Induced Parkinson' Model. *Crystals* **2022**, *12*, 1518. [CrossRef]
16. Wang, Y.; Xu, Y.; Zheng, Z.; Xue, M.; Meng, Z.; Xu, Z.; Li, J.; Lin, Q. Studies on the Crystal Forms of Istradefylline: Structure, Solubility and Dissolution Profile. *Crystals* **2022**, *12*, 917. [CrossRef]
17. Pérez-Aguilar, C.D.; Islas, S.R.; Moreno, A.; Cuéllar-Cruz, M. The Effect of DNA from Escherichia Coli at High and Low CO_2 Concentrations on the Shape and Form of Crystalline Silica-Carbonates of Barium (II). *Crystals* **2022**, *12*, 1147. [CrossRef]
18. Tanaka, H.; Utata, R.; Tsuganezawa, K.; Takahashi, S.; Tanaka, A. Through Diffusion Measurements of Molecules to a Numerical Model for Protein Crystallization in Viscous Polyethylene Glycol Solution. *Crystals* **2022**, *12*, 881. [CrossRef]
19. Neron, V.; Gelin, P.; Hashemiesfahan, M.; De Malsche, M.; Lutsko, J.F.; Maes, D.; Galand, Q. The effect of Controlled Mixing on ROY Polymorphism. *Crystals* **2022**, *12*, 577.
20. Radel, B.; Gleiβ, M.; Nirschl, H. Crystal Breakage Due to Combined Normal and Shear Loading. *Crystals* **2022**, *12*, 644. [CrossRef]
21. Kubiak, M.; Kampen, I.; Schilde, C. Structure-Based Modeling of Mechanical Behaviour of Cross-Linked Enzyme Crystals. *Crystals* **2022**, *12*, 441. [CrossRef]
22. Ward, A.R.; Dmytriw, S.; Vajapayajula, A.; Snow, C.D. Stabilizing DNA-Protein Co-Crystals via Intra-Crystal chemical Ligation of the DNA. *Crystals* **2022**, *12*, 49. [CrossRef]
23. Martínez-Rodríguez, S.; Contreras-Montoya, R.; Torres, J.M.; Álvarez de Cienfuegos, L.; Gavira, J.A. A New L-Proline Amide Hydrolase with Potential Application within the Amidase Process. *Crystals* **2022**, *12*, 18. [CrossRef]
24. Radic', Z. Shifts in Backbone Conformation of Acetylcholinesterase upon Binding of Covalent Inhibitors, Reversible Ligands and Substrates. *Crystals* **2022**, *11*, 1557. [CrossRef]

MDPI

Article

Synthesis, Molecular Docking, and Neuroprotective Effect of 2-Methylcinnamic Acid Amide in 1-methyl-4-phenyl-1,2,3,6-tetrahydropyridine (MPTP)—An Induced Parkinson's Disease Model

Maya Chochkova [1,*], Rusi Rusew [2], Reni Kalfin [3], Lyubka Tancheva [3], Maria Lazarova [3], Hristina Sbirkova-Dimitrova [2], Andrey Popatanasov [3], Krasimira Tasheva [4], Boris Shivachev [2], Nejc Petek [5] and Martin Štícha [6]

1 Faculty of Mathematics and Natural Sciences, South-West University "Neofit Rilski", 66 Ivan Mihailov Str., 2700 Blagoevgrad, Bulgaria
2 Institute of Mineralogy and Crystallography "Akad. Ivan Kostov", Bulgarian Academy of Sciences, Acad. G. Bonchev bl., 107 Sofia, Bulgaria
3 Institute of Neurobiology, Bulgarian Academy of Sciences, Acad. G. Bonchev Bl.23, 1113 Sofia, Bulgaria
4 Institute of Plant Physiology and Genetics, Bulgarian Academy of Sciences, 1113 Sofia, Bulgaria
5 Faculty of Chemistry and Chemical Technology, University of Ljubljana, Večna pot 113, 1000 Ljubljana, Slovenia
6 Section of Chemistry, Faculty of Science, Charles University, Hlavova 2030/8, 12843 Prague, Czech Republic
* Correspondence: mayabg2002@yahoo.com

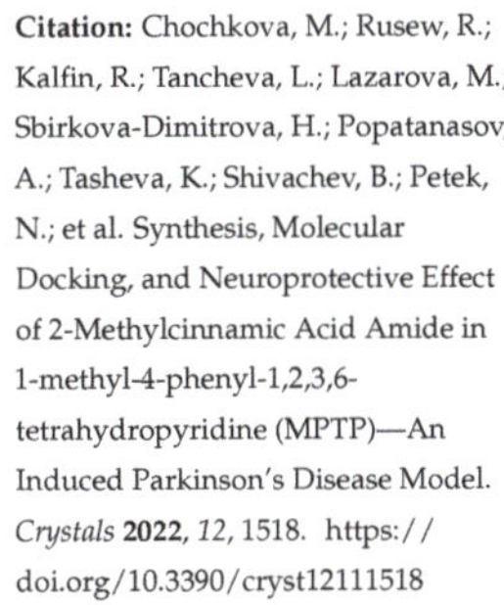

Citation: Chochkova, M.; Rusew, R.; Kalfin, R.; Tancheva, L.; Lazarova, M.; Sbirkova-Dimitrova, H.; Popatanasov, A.; Tasheva, K.; Shivachev, B.; Petek, N.; et al. Synthesis, Molecular Docking, and Neuroprotective Effect of 2-Methylcinnamic Acid Amide in 1-methyl-4-phenyl-1,2,3,6-tetrahydropyridine (MPTP)—An Induced Parkinson's Disease Model. *Crystals* **2022**, *12*, 1518. https://doi.org/10.3390/cryst12111518

Academic Editor: Abel Moreno

Received: 8 September 2022
Accepted: 21 October 2022
Published: 26 October 2022

Abstract: Parkinson's disease (PD) has emerged as the second most common form of human neurodegenerative disorders. However, due to the severe side effects of the current antiparkinsonian drugs, the design of novel and safe compounds is a hot topic amongst the medicinal chemistry community. Herein, a convenient peptide method, TBTU (O-(benzotriazole-1-yl)-N,N,N′,N′-tetramethyluronium tetrafluoroborate), was used for the synthesis of the amide (*E*)-*N*-(2-methylcinnamoyl)-amantadine (CA(2-Me)-Am; 3)) derived from amantadine and 2-methylcinnamic acid. The obtained hybrid was studied for its antiparkinsonian activity in an experimental model of PD induced by MPTP. Mice (C57BL/6,male, 8 weeks old) were divided into four groups as follows: (1) the control, treated with normal saline (i.p.) for 12 consecutive days; (2) MPTP (30 mg/kg/day, i.p.), applied daily for 5 consecutive days; (3) MPTP + CA(2-Me)-Am, applied for 12 consecutive days, 5 days simultaneously with MPTP and 7 days after MPTP; (4) CA(2-Me)-Am +oleanoic acid (OA), applied daily for 12 consecutive days. Neurobehavioral parameters in all experimental groups of mice were evaluated by rotarod test and passive avoidance test. Our experimental data showed that CA(2-Me)-Am in parkinsonian mice significantly restored memory performance, while neuromuscular coordination approached the control level, indicating the ameliorating effects of the new compound. In conclusion, the newly synthesized hybrid might be a promising agent for treating motor disturbances and cognitive impairment in experimental PD.

Keywords: amantadine; 2-methylcinnamic acid; single crystal X-ray diffraction; Parkinson's disease

1. Introduction

Globally, the growth of the older population, comprising 7 % or more of the total population, is projected to reach about 2 billion people by 2050 [1]. Thereafter, this can be associated with increasing prevalence of age-related diseases as neurodegenerative diseases, such as Parkinson's disease and Alzheimer's disease, among others. According to the WHO (World Health Organization), amongst the neurological disorders, there is a growing concern around the disability and death caused by PD [2]. Parkinsonism is the second most common neurodegenerative disaster after Alzheimer's disease [3,4]. Currently, there are no strategies that can stop the brain cell injury afflicted by PD. The multifactorial nature of this incurable pathology requires an effective multi-target concept that can hit diverse targets. However, the almost all of the central nervous system drug candidates do not efficiently penetrate the blood–brain barrier. Therefore, to solve this problem, it is wildly accepted that the addition of a lipophilic rest to main structure can modify absorption, distribution, metabolism, or excretion (ADME) properties of an entire molecule.

Accordingly, adamantane core has been known as a precise building block that can alter the lipophilicity on a lead compound, without increasing its toxicity. Indeed, there are many adamantane-based compounds that are currently used in clinical practice and also as potential therapeutics [5,6]. However, the end of the era of aminoadamantanes, including rimantadine and amantadine as antiviral, is delineated, since they have faced increasing resistance against influenza A strains [7–9]. Surprisingly, a random finding resurrected the role of amantadine in the efficacy for symptomatic alleviation in PD [10], as well as for other movement disorders [11]. Additionally, the dual effects of amantadine on parkinsonian signs and symptoms and levodopa-induced dyskinesias are due to its dopaminergic and glutamatergic properties [11]. Amantadine is the only glutamate antagonist drug that is prescribed against PD, often used to treat dyskinesia. However, the clinical use of amantadine is limited because of concerns regarding its safety and tolerability issues, as well as the duration of its antidyskinetic efficacy. Hence, the search for new agents with powerful antiparkinsonian action and good biological tolerance is an important task for chemists and biologists.

In this regard, cinnamic acid (CA) has been known as a plant 3-phenylpropenoic acid and represents one of the constituents in the common spice cinnamon. This unsaturated carboxylic acid has been obtained through phenylpropanoid pathway as a deaminated plant product of its amino acid precursor phenylalanine. Besides having a plethora of activities such as antidiabetic [12] and anti-cancer [13] effects, cinnamic acid emerges with a new function in protecting dopaminergic neurons via PPARα [14]. Moreover, an earlier result [15] reveals that cinnamic acid improves memory by suppressing the oxidative stress and cholinergic dysfunction in the brain of diabetic mice.

Inspired by the above-mentioned results for CA, in our study, we report the synthesis of a hybrid molecule consisting of methylated cinnamic acid and amantadine. Furthermore, the newly obtained derivative was examined as a potential neuroprotective agent in the MPTP experimental mouse model of PD.

2. Materials and Methods

2.1. General Methods

2-Methyl-cinnamic acid, amantadine, and other reagents were purchased from Angene Chemical, Sigma Aldrich (FOT, Bulgaria), whereas all solvents were obtained from Thermo Fisher Scientific (Bulgaria) and applied with no further purification. Thin-layer chromatography (TLC) was carried out on precoated Kieselgel $60F_{254}$ plates (Merck, Germany) with detection by UV absorbance at 254 nm. A TLC plate was visualized by Ce-PMo reagent solution followed by heating. Flash chromatography of the target amide was performed on prepackaged BÜCHI FlashPure EcoFlex silica columns.

The newly amide 3 was synthesized according to the modified literature method [16]. The NMR spectra were recorded in deuterated solvents with $(CH_3)_4Si$ as the internal standard on a Bruker Ascend neo NMR 600 instrument (Bruker, Billerica, MA, USA) at 600 MHz for ^1H nuclei and at 151 MHz for ^{13}C nuclei. A Bruker Compact QTOF-MS (Bruker Daltonics, Bremen, Germany) controlled by the Compass 1.9 Control software was used to measure the mass spectrum. The monoisotopic mass values were calculated using Data analysis software v 4.4 (Bruker Daltonics, Germany). The analysis was conducted in the positive ion mode at a scan range from m/z 50 to 1000, and nitrogen was used as nebulizer gas at a pressure of 4 psi and flow of 3 L/min for the dry gas. The capillary voltage and temperature were set at 4500 V and 220 °C, respectively.

2.2. Synthesis of (E)- N-(2-Methylcinnamoyl)-Amantadine (3)

2-Methylcinnamic acid (1.8 g, 11.4 mmol) was suspended in 30 mL of CH_2Cl_2, and then, after adding Et_3N (1.6 mL, 11.4 mmol), the obtained colorless liquid was treated by solid TBTU (3.7 g, 11.4 mmol). After being stirred for ≈10 min, to the mixture, we added amantadine (2.4 g, 12.6 mmol) and Et_3N (1.8 mL, 12.6 mmol), dissolved (under sonication) in 40 mL CH_2Cl_2. Thus, the reaction mixture was stirred at room temperature for 3 h, and then was diluted with an additional 30 mL CH_2Cl_2. The organic phase was washed with 5% aqueous $NaHCO_3$ (5 × 50 mL) and brine (3 × 50 mL), dried over Na_2SO_4, and concentrated in vacuo. Furthermore, after purification, the amide was obtained (3.3 g, 89%) as white crystals.

Compound (**3**): white crystals (CH_3CN); mp 188–189 ^{0}C; ^{1}H NMR (DMSO-d_6, 600 MHz) δ 7.59 (s, 1H), 7.56 (d, J = 15.6 Hz, 1H), 7.48 (d, J = 7.3 Hz, 1H), 7.27–7.20 (m, 3H), 6.58 (d, J = 15.6 Hz, 1H), 2.35 (s, 3H), 2.03 (bs, 3H), 2.00 (bs, 6H), 1.64 (bs, 6H); ^{13}C NMR (DMSO-d_6, 151 MHz) δ 164.5, 137.0, 135.7, 134.4, 131.1, 129.4, 126.8, 126.3, 125.3, 51.4, 41.5, 36.5, 29.3, 19.9; HRMS m/z 318.1830 (calcd for $C_{20}H_{25}NNaO$, 318.1828).

2.3. Single Crystal X-ray Diffraction (SCXRD) of (E)-N-(2-methylcinnamoyl)-amantadine (3)

Single crystals of compound 3 were grown from 1:1 *v/v* benzene methanol solution. A crystal with suitable size and quality was selected and was mounted on a glass capillary. The diffraction peak intensities and coordinates were collected on Bruker D8 Venture diffractometer (Bruker AXS GmbH, Karlsruhe, Germany) equipped with a PhotonII CMOS detector using micro-focus Mo*K*α radiation (λ = 0.71073 Å). Data were processed with CrysAlisPro software [17]. The structure was solved with intrinsic methods using ShelxT [18] and refined by the full-matrix least-squares method on the F^2 with ShelxL program [19]. All non-hydrogen atoms were located successfully from the Fourier map and were refined anisotropically. Hydrogen atoms were placed on calculated positions (C–$H_{aromatic}$ = 0.93, C–H_{methyl} = 0.96 Å, and C–$H_{methylenic}$ = 0.97 Å, riding on the parent atom (*Ueq* = 1.2). The H atom near the nitrogen was located from a different Fourier map. Complete crystallographic data for the structure of the title compound reported in this paper were deposited in the CIF format at the Cambridge Crystallographic Data Center as 2205297. These data can be obtained free of charge via http://www.ccdc.cam.ac.uk/conts/retrieving.html, deposited on 5 September 2022 (or from the CCDC, 12 Union Road, Cambridge CB2 1EZ, UK; Fax: +441223336033; e-mail: deposit@ccdc.cam.ac.uk).

2.4. Neurobehavioral Studies

Mice (C57BL/6, male, 8 weeks old) were obtained from Erboj (Animal Breeding Center, Slivniza, Sofia). The animals were housed two per cage under constant laboratory conditions (25 ± 3°C, 12/12 h light/dark cycle) with food and water available ad libitum. The habituation period was 5 days before the start of the experiment. The protocol of all experiments was in accordance with the requirements of the European Communities Council Directive 86/609/EEC and rules of the Bioethics Committee 30/03/2021, Institute of Neurobiology, Bulgarian Academy of Sciences.

CA(2-Me)-Am (3) was dissolved in oleanolic acid (OA). We tested five doses of amide 3 applied per os on 25 male mice C57BL/6 and found the dose of 20 mg/kg to be the most effective.

Mice were divided into four experimental groups (*n* = 8 in each group) as follows: (1) control, treated with normal saline (i.p.) for 12 consecutive days; (2) MPTP (30 mg/kg/day, i.p.) applied daily for 5 consecutive days in accordance with the work of Shin et al. [20]; (3) MPTP (30 mg/kg, i.p.) + CA(2-Me)-Am (20 mg/kg, per os) applied for 12 consecutive days, 5 days simultaneously with MPTP and 7 days after MPTP; (4) MPTP (30 mg/kg, i.p.) + OA (per os) applied daily for 12 consecutive days.

All mice training was conducted before MPTP administration.

2.4.1. Rotarod Test

Mice from all experimental groups were placed on a gyratory with a fixed speed of 7 rpm/min, and the time on rotarod was determined. The observation period was 5 min. All animals were pre-trained on the rotarod apparatus before treatment in order to reach stable performance. The training consisted of one session per day over 3 consecutive days. The test was made on the 13th day, and the average time per group was calculated after the experiment and repeated four times [21].

2.4.2. Passive Avoidance Test

Learning and memory performance in mice was evaluated using passive avoidance learning test [22]. Acquisition phase: during this phase, each animal was placed in the illuminated compartment. When the rodents innately entered into the dark compartment, they received a mild electrical foot shock (0.5 mA, 3 s). In this trial, the initial latency (IL) of entrance into the dark chamber of each animal was recorded, and mice with ILs > 60 s were excluded from the study. Test phase: on the 13th and 14th days, each mouse was placed in the illuminated chamber, and the entry into the dark chamber was measured as step through latency (STL). The behavioral observations were carried out between 9 a.m. and 12 a.m.

2.4.3. Statistical Analysis

The results were expressed as means ± the standard error of the mean (SEM) or as percentage changes over the mean compared to the control. Statistical analyses of the data were performed by one-way analysis of variance (ANOVA) followed by Dunnett post hoc comparison test. Differences were considered significant at $p < 0.05$.

3. Results and Discussion

3.1. Chemistry

Herein, the 2-methylcinnamic acid amide (CA(2-Me)-Am; 3) was synthesized as outlined in Scheme 1. Generally, the amidation of 2-methylcinnamic acid (CA(2-Me)-OH; 1) with amantadine (Am; 2) was carried out in the presence of tertiary amine (triethylamine, Et_3N) and by one of the preferred coupling reagents for in situ activation, such as TBTU [16] to amide 3.

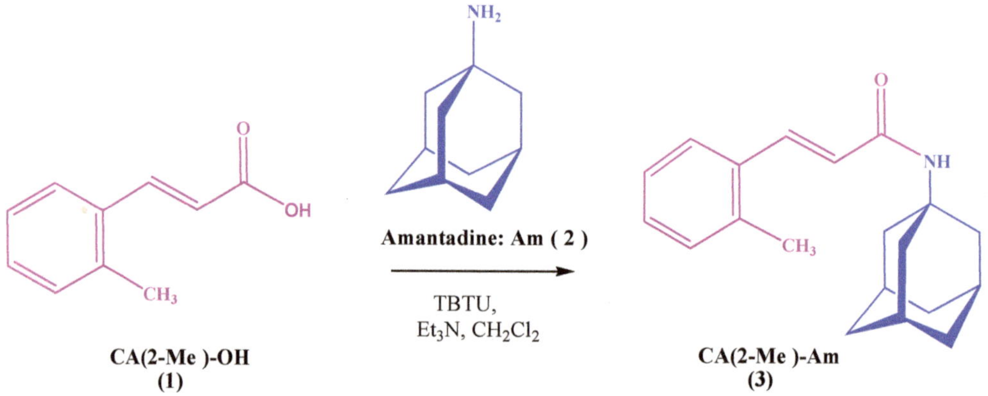

NH2
O
OH
CH3
CA(2-Me)-OH
(1)
Amantadine: Am (2)
TBTU,
Et_3N, CH_2Cl_2
O
NH
CH3
CA(2-Me)-Am
(3)

Scheme 1. Synthetic route of amide **3**.

The structure of the newly obtained compound (3) was confirmed by the ^{1}H NMR, ^{13}C NMR, HRMS, and single crystal analysis powder diffraction. The title compound (CA(2-Me)-Am; 3) crystallized in the monoclinic $P2_1/c$ space group, with one molecule in the asymmetric unit (Figure 1). The bond distances and angles (Table 1) within the adamantane and 2-methylcinnamic acid were comparable with those observed in other structures [23–27].

Table 1. The most important data collection and crystallographic refinement parameters for (E)=N-(2-methylcinnamoyl)-amantadine.

Empirical formula	$C_{20}H_{25}NO$
Formula weight	295.41
Temperature/K	290.00
Crystal system	monoclinic
Space group	$P2_1/c$
a/Å	14.692(2)
b/Å	11.900(2)
c/Å	9.9904(18)
α/°	90
β/°	104.943(5)
γ/°	90
Volume/Å^3	1687.6(5)
Z	4
ρ_{calc}g/cm^3	1.163
μ/mm^{-1}	0.071
F(000)	640.0
Crystal size/mm^3	0.3 × 0.25 × 0.2
Radiation	MoKα (λ = 0.71073)
2Θ range for data collection/°	4.466 to 52.842
Index ranges	$-18 \leq h \leq 17$, $-14 \leq k \leq 12$, $-12 \leq l \leq 12$
Reflections collected	10,910
Independent reflections	3443 (R_{int} = 0.0595, R_{sigma} = 0.0654)
Data/restraints/parameters	3443/0/205
Goodness-of-fit on F^2	1.018
Final R indexes (I > =2σ (I))	R_1 = 0.0605, wR_2 = 0.1171
Final R indexes (all data)	R_1 = 0.1142, wR_2 = 0.1405
Largest diff. peak/hole/e Å^{-3}	0.17/−0.13

The angle between the phenyl and acrylamide moieties was 29.8 °, disclosing that the conjugation was not stringent, e.g., the conjugation could be disrupted. In the molecule of (*E*)–*N*-(2-methylcinnamoyl)-amantadine, one hydrogen donor (N–H) and one acceptor (carbonyl oxygen) were present. In the crystal structure, the molecules produced one-dimensional chains with a graph set $C_1^1(4)$ [28,29] (Figure 2a). The three-dimensional packing of the molecules (Figure 2b) did not reveal additional weak interactions, and thus the stabilization of the crystal structure was achieved by the N1-H1 . . . O1 hydrogen bond (N1 . . . O1 of 3.065(5) Å).

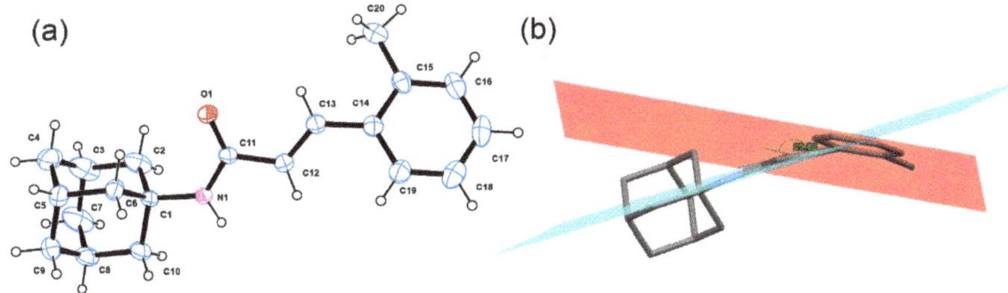

Figure 1. (**a**) ORTEP [30] view and numbering scheme of the molecule present in the asymmetric unit of (*E*)–*N*-(2-methylcinnamoyl)-amantadine; the thermal ellipsoids were drawn with 50% probability, and hydrogen atoms are shown as small spheres with arbitrary radii. (**b**) Observed angle between the mean plane of the phenyl (C1/C15/C16/C17/C18/C19) and acrylamide (C13/C12/C11/O1/N1) moieties (C13/C12/C11/O1/N1).

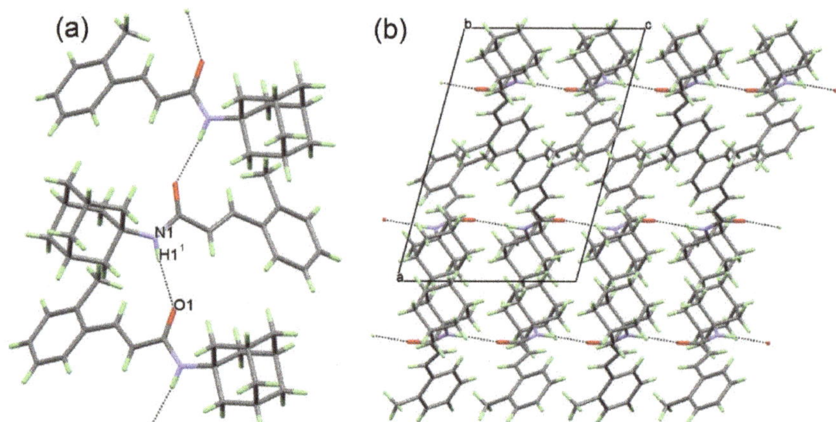

Figure 2. The observed (**a**) hydrogen bonding interaction stabilizing the crystal structure of (*E*)–*N*-(2-methylcinnamoyl)-amantadine and (**b**) a view along the *b* axis of the three-dimensional packing of the molecules and formation of $C_1^1(4)$ chains propagating along the [10] plane.

3.2. In Vivo Evaluation of Amide 3 in an Experimental Mouse Model of PD

3.2.1. Effect of CA(2-Me)-Am (3) on the Weight of Experimental Animals

There was no significant change in the weight of the control mice over the 12-day period. In those treated with MPTP, we observed a 6.92% weight gain within the group. In the MPTP + CA(2-Me)-Am and MPTP + OA mice, weight reduction was recorded at the end of the observed period at 17.41% and 15.51%, respectively (Figure 3).

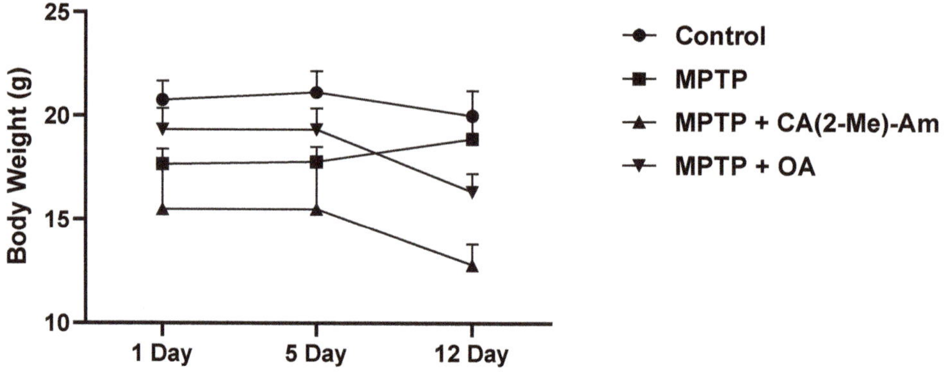

Figure 3. Effect of CA(2-Me)-Am on the weight of the experimental animals. Data are presented as means with their respective standard errors (m ± S.E.M; n = 8; * $p < 0.05$).

3.2.2. Rotarod Test

The studies performed demonstrated that the group of mice treated either with the MPTP toxin or with MPTP + OA spent less time on the rotating lever of the rotarod apparatus as compared to the controls, which is an indication of a motor-impairing effect. The reduction was by 21.13 % ($p < 0.05$) for the MPTP group, and by 20.99 % ($p < 0.05$) for the MPTP + OA group (Figure 4). In the MPTP + CA(2-Me)-Am group, the time that experimental animals spent on the rotary lever was comparable to that of the control group (Figure 4).

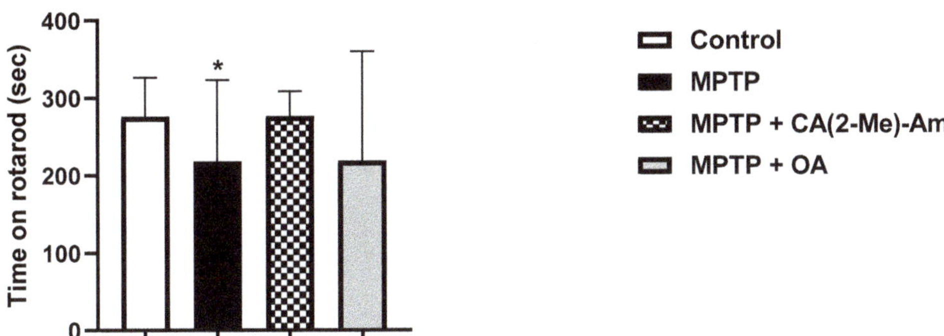

Figure 4. Effect of CA(2-Me)-Am on neuromuscular coordination. The asterisks above bars indicate significant differences in number of falls per minute for each experimental group versus the control at * $p < 0.05$. Statistical analysis was performed by one-way analysis of variance (ANOVA) followed by Dunnett's post hoc comparison test.

3.2.3. Passive Avoidance Test

The administration of the MPTP toxin caused a decrease in the step-through latency time by 31.56% ($p < 0.01$) at 1st h and by 33.45% ($p < 0.01$) at 24th h after the training of mice as compared to controls, which is evidence of memory and learning deficits. Administration of CA(2-Me)-Am increased the latent reaction time by 33.49% ($p < 0.05$) at the 1st h and by 33.84% ($p < 0.05$) at 24th h as compared to the MPTP-treated group, an indication of a memory-protective effect of the newly synthesized amantadine derivative (Figure 5).

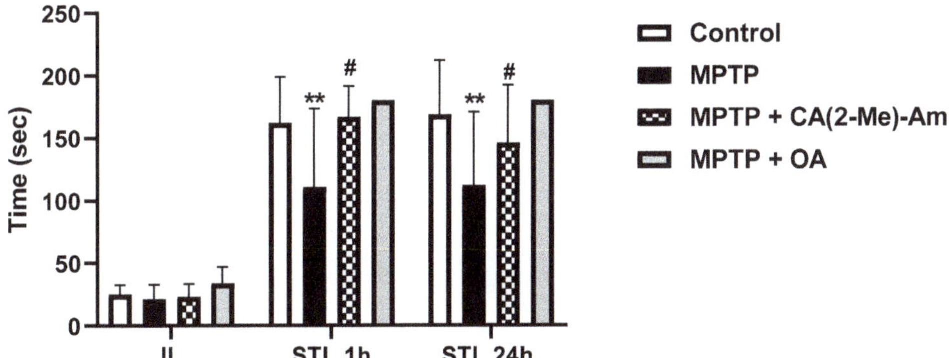

Figure 5. The effect of CA(2-Me)-Am on initial latency (IL) and step-through latency (STL) in a single-trial passive avoidance test in a mouse model of PD. Significance vs. control group: **$p < 0.01$; significance vs. MPTP-treated group: # $p < 0.05$. Statistical analysis was performed by one-way analysis of variance (ANOVA), followed by Dunnett's post hoc comparison test.

3.3. Molecular Docking

The docking of (*E*)-*N*-(2-methylcinnamoyl)-amantadine (**3**) was performed against four different targets associated with PD: A_{2a}AR (3EML) [31], COMT (1H1D) [32], MAO-B (2C65) [33], and NMDA (7SAD) [34] (Table 2).

Table 2. Molecular docking score (kcal/mol) of (*E*)–*N*-(2-methylcinnamoyl)-amantadine against selected targets associated with PD.

	Molegro Virtual Docker Score	Detected Hydrogen Bonding Interaction
	(*E*)–*N*-(2-methylcinnamoyl)-amantadine	
NMAD (7SAD) [34]	-80.240	No
COMT (1H1D) [32]	-111.957	No
A2aAR (3EML) [31]	-83.578	C=O ... O-H Tyr271 D ... A 3.38 Å
MAO-B (2C65) [33]	-119.889	C=O ... N-H Gly58 D ... A 3.09 Å

NMAD, *N*-methyl-*D*-aspartate receptor; COMT, catechol-*O*-methyltransferase; A2aAR, A2A adenosine receptor; MAO-B, monoamine oxidase B.

The docking approach involved predicting the conformation and orientation of ligands within a targeted binding site. Initially, the reference drug memantine present in 7SAD [34] was employed to adjust the docking parameters. The structures of the target enzymes were obtained from the Protein Data Bank (PDB), the coordinates of the small molecules were generated from the crystal structure of the (*E*)–*N*-(2-methylcinnamoyl)-amantadine, and positioning in the active site and docking were conducted using Molegro Virtual Docker (MVD2019.7.0.0-2019-03-18-1B win32). Details regarding the docking validation are provided in Figures S4 and S5. The UCSF Chimera [35] and Ligplot+ 2.2.5 [36] were used for visualization and interactions detection. On the basis of the scores obtained from the docking results, the interaction of (*E*)–*N*-(2-methylcinnamoyl)-amantadine with MAO-B was most favorable (score of –119.889, Figure 6). The second possibility not to be excluded is the interaction with COMT (–111.957). However, while for MAO-B, a hydrogen bonding interaction was detected for COMT, no hydrogen bonding interaction "enzyme ... ligand" was identified (Figure 7). Interestingly, the active sites of COMT and MAO-B shared a large amount of analogical AA and thus it may be possible to design a ligand that will interact with both enzymes.

Parkinson's disease has high social significance, resulting from a progressive loss of nigrostriatal dopaminergic neurons. The decrease in striatal dopaminergic innervation due to this loss is responsible for motor disturbances characteristic of the disease, such as akinesia, muscular rigidity, and tremor, and cognitive function impairment later appears. Amantadine is an agent that raises the concentration of dopamine in the synaptic cleft in PD. Currently, levodopa is considered to be the gold standard for symptomatic treatment of PD. However, long-term treatment with levodopa is complicated by motor fluctuations and dyskinesia. Everything stated above requires the search for compounds that can replace levodopa and improve the efficacy of amantadine in the treatment of PD. In this line of thinking, we performed investigations of a newly obtained amantadine derivative CA(2-Me)-Am (**3**) on neuromuscular coordination, learning, and memory in an experimental mouse model of PD. The obtained data showed that amide 3 restored neuromuscular coordination and memory performance of parkinsonian animals to the control level, giving an indication of its beneficial protective effects.

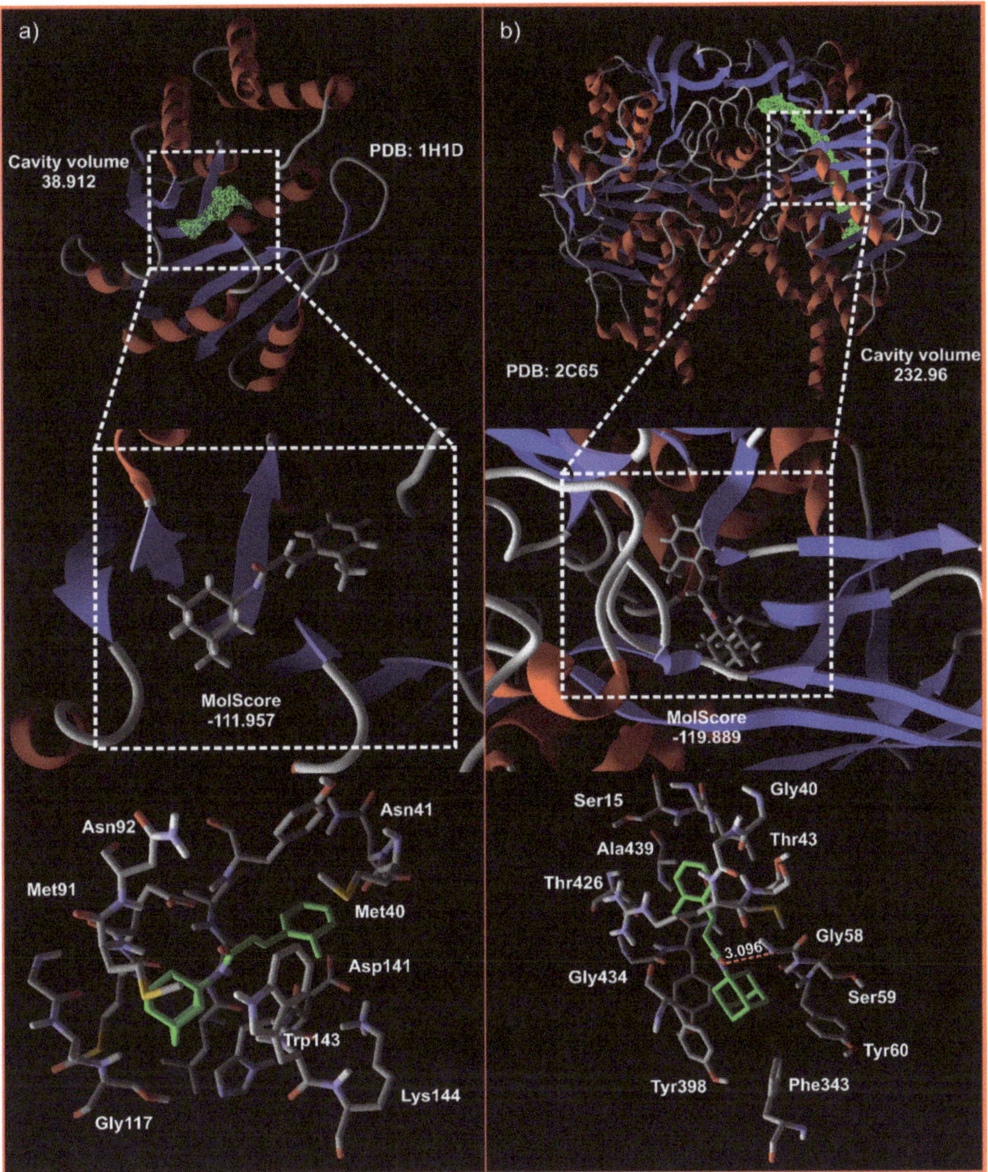

Figure 6. Visualization of the docking studies and molecular interaction of (*E*)-*N*-(2-methylcinnamoyl)-amantadine with (**a**) COMT and (**b**) MAO-B. The hydrogen bonding interaction *N*-H . . . O=C is shown in as red dashed line with the A . . . D distance of 3.096 Å.

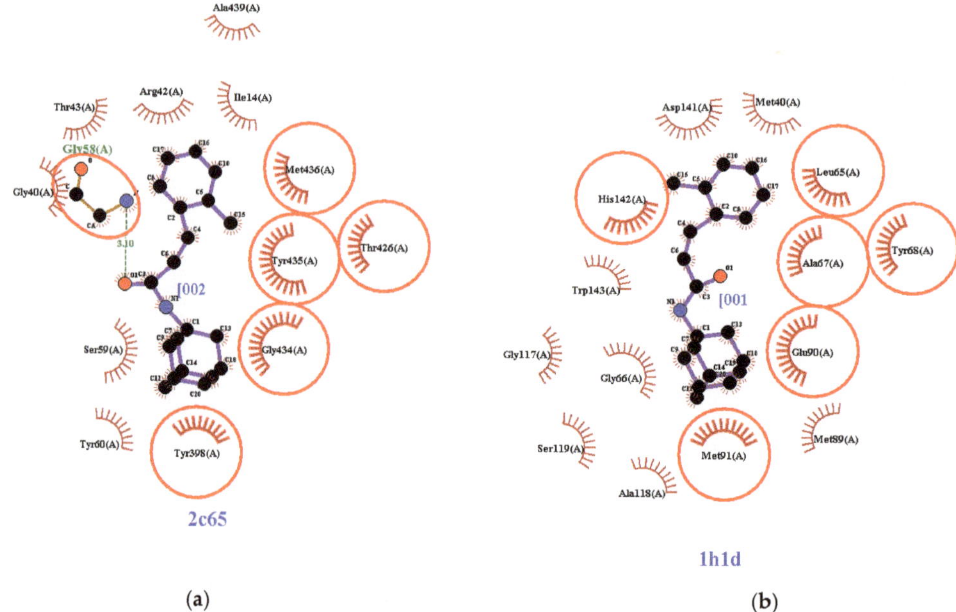

Figure 7. Observed interactions after docking of (*E*)–*N*-(2-methylcinnamoyl)-amantadine into the active site of two putative PD targets (**a**) MAO-B (2C65) and (**b**) COMT (1H1D). The hydrogen bonding interactions are shown in green; the hydrophobic contacts are shown as [symbol] for enzymes and with ● for ligands; the similar residues for both enzymes are shown as [symbol].

Moreover, the molecular docking study investigations showed that from the considered four targets, (*E*)–*N*-(2-methylcinnamoyl)-amantadine interacted preferably with MAO-B, followed by COMT. The detected hydrogen bonding interaction could be used for development of modified potential antiviral drug candidates.

In conclusion, our results demonstrated ameliorating effects of the newly synthesized compound CA(2-Me)-Am in an experimental model of Parkinson's disease, which deserves further investigations.

Supplementary Materials: The following supporting information can be downloaded at: https://www.mdpi.com/article/10.3390/cryst12111518/s1, NMR and MS spectra of compound 3 are provided.

Author Contributions: Synthesis, supervision, writing and manuscript conceptualization, M.C.; NMR spectroscopy studies, N.P. and M.Š.; study, M.Š.; single crystal X-ray diffraction experiments, R.R. and H.S.-D.; docking studies, writing, B.S.; neurobehavioral studies, writing, R.K., L.T., M.L., A.P. and K.T. All authors have read and agreed to the published version of the manuscript.

Funding: This work was funded by the Bulgarian National Science Fund (BNSF), grant number KP-06-Russia/7-2019.

Institutional Review Board Statement: The animal study protocol was approved by the Commission of Bioethics (CBE) at the Institute of Neurobiology, Bulgarian Academy of Sciences—BAS/CBE/018/2020.

Informed Consent Statement: Not applicable.

Data Availability Statement: Crystallographic data for the structure of the title compound *(E)*-*N*-(2-methylcinnamoyl)-amantadine (CA(2-Me)-Am; 3)) was deposited in the CIF format with the Cambridge Crystallographic Data Center as 2205297. These data can be obtained free of charge via http://www.ccdc.cam.ac.uk/conts/retrieving.html, deposited on 05 September 2022 (or from the CCDC, 12 Union Road, Cambridge CB2 1EZ, UK; Fax: +441223336033; e-mail: deposit@ccdc.cam.ac.uk).

Conflicts of Interest: The authors declare no conflict of interest.

References

1. He, W.; Goodkind, D.; Kowal, P.R. *An Aging World: 2015*; International Population Reports; United States Census Bureau: Suitland-Silver Hill, MD, USA, 2016.
2. *Parkinson Disease: A Public Health Approach: Technical Brief*; World Health Organization: Geneva, Switzerland, 2022.
3. Mhyre, T.R.; Boyd, J.T.; Hamill, R.W.; Maguire-Zeiss, K.A. Parkinson's Disease. *Subcell. Biochem.* **2012**, *65*, 389–455. [CrossRef] [PubMed]
4. Wyss-Coray, T. Ageing, Neurodegeneration and Brain Rejuvenation. *Nature* **2016**, *539*, 180–186. [CrossRef] [PubMed]
5. Liu, J.; Obando, D.; Liao, V.; Lifa, T.; Codd, R. The Many Faces of the Adamantyl Group in Drug Design. *Eur. J. Med. Chem.* **2011**, *46*, 1949–1963. [CrossRef] [PubMed]
6. Wanka, L.; Iqbal, K.; Schreiner, P.R. The Lipophilic Bullet Hits the Targets: Medicinal Chemistry of Adamantane Derivatives. *Chem. Rev.* **2013**, *113*, 3516–3604. [CrossRef] [PubMed]
7. Schmidtke, M.; Zell, R.; Bauer, K.; Krumbholz, A.; Schrader, C.; Suess, J.; Wutzler, P. Amantadine Resistance among Porcine H1N1, H1N2, and H3N2 Influenza A Viruses Isolated in Germany between 1981 and 2001. *Intervirology* **2006**, *49*, 286–293. [CrossRef]
8. Nelson, M.I.; Simonsen, L.; Viboud, C.; Miller, M.A.; Holmes, E.C. The Origin and Global Emergence of Adamantane Resistant A/H3N2 Influenza Viruses. *Virology* **2009**, *388*, 270–278. [CrossRef]
9. Weinstock, D.M.; Zuccotti, G. The Evolution of Influenza Resistance and Treatment. *JAMA* **2009**, *301*, 1066–1069. [CrossRef]
10. Schwab, R.S.; England, A.C.; Poskanzer, D.C.; Young, R.R. Amantadine in the Treatment of Parkinson's Disease. *JAMA* **1969**, *208*, 1168–1170. [CrossRef]
11. Rascol, O.; Fabbri, M.; Poewe, W. Amantadine in the Treatment of Parkinson's Disease and Other Movement Disorders. *Lancet Neurol.* **2021**, *20*, 1048–1056. [CrossRef]
12. Adisakwattana, S. Cinnamic Acid and Its Derivatives: Mechanisms for Prevention and Management of Diabetes and Its Complications. *Nutrients* **2017**, *9*, E163. [CrossRef]
13. Liu, L.; Hudgins, W.R.; Shack, S.; Yin, M.Q.; Samid, D. Cinnamic Acid: A Natural Product with Potential Use in Cancer Intervention. *Int. J. Cancer* **1995**, *62*, 345–350. [CrossRef]
14. Prorok, T.; Jana, M.; Patel, D.; Pahan, K. Cinnamic Acid Protects the Nigrostriatum in a Mouse Model of Parkinson's Disease via Peroxisome Proliferator-Activated Receptorα. *Neurochem. Res.* **2019**, *44*, 751–762. [CrossRef]
15. Hemmati, A.A.; Alboghobeish, S.; Ahangarpour, A. Effects of Cinnamic Acid on Memory Deficits and Brain Oxidative Stress in Streptozotocin-Induced Diabetic Mice. *Korean J. Physiol. Pharmacol. Off. J. Korean Physiol. Soc. Korean Soc. Pharmacol.* **2018**, *22*, 257–267. [CrossRef]
16. Knorr, R.; Trzeciak, A.; Bannwarth, W.; Gillessen, D. New Coupling Reagents in Peptide Chemistry. *Tetrahedron Lett.* **1989**, *30*, 1927–1930. [CrossRef]
17. Rigaku Oxford Diffraction. CrysAlis pro. Rigaku Oxford Diffraction, CrysAlis pro CrysAlis Pro. 2015. Available online: https://www.rigaku.com/products/crystallography/crysalis (accessed on 20 October 2022).
18. Sheldrick, G.M. SHELXT-Integrated Space-Group and Crystal-Structure Determination. *Acta Crystallogr. Sect. Found. Adv.* **2015**, *71*, 3–8. [CrossRef]
19. Sheldrick, G.M. A Short History of SHELX. *Acta Crystallogr. A* **2008**, *64*, 112–122. [CrossRef]
20. Shin, K.S.; Zhao, T.T.; Choi, H.S.; Hwang, B.Y.; Lee, C.K.; Lee, M.K. Effects of Gypenosides on Anxiety Disorders in MPTP-Lesioned Mouse Model of Parkinson's Disease. *Brain Res.* **2014**, *1567*, 57–65. [CrossRef]
21. Manna, S.; Bhattacharyya, D.; Mandal, T.K.; Dey, S. Neuropharmacological Effects of Deltamethrin in Rats. *J. Vet. Sci.* **2006**, *7*, 133–136. [CrossRef]
22. Shahidi, S.; Komaki, A.; Mahmoodi, M.; Atrvash, N.; Ghodrati, M. Ascorbic Acid Supplementation Could Affect Passive Avoidance Learning and Memory in Rat. *Brain Res. Bull.* **2008**, *76*, 109–113. [CrossRef]
23. Land, M.A.; Robertson, K.N.; Ylijoki, K.E.O.; Clyburne, J.A.C. Reactivity of 1,3-Dichloro-1,3-Bis(Dimethylamino)-Propenium Salts with Primary Amines. *New J. Chem.* **2021**, *45*, 13558–13570. [CrossRef]
24. Pereira, A.K.d.S.; Manzano, C.M.; Nakahata, D.H.; Clavijo, J.C.T.; Pereira, D.H.; Lustri, W.R.; Corbi, P.P. Synthesis, Crystal Structures, DFT Studies, Antibacterial Assays and Interaction Assessments with Biomolecules of New Platinum (II) Complexes with Adamantane Derivatives. *New J. Chem.* **2020**, *44*, 11546–11556. [CrossRef]
25. Perlovich, G.L.; Ryzhakov, A.M.; Tkachev, V.V.; Hansen, L.K.; Raevsky, O.A. Sulfonamide Molecular Crystals: Structure, Sublimation Thermodynamic Characteristics, Molecular Packing, Hydrogen Bonds Networks. *Cryst. Growth Des.* **2013**, *13*, 4002–4016. [CrossRef]

26. Sarcevica, I.; Orola, L.; Veidis, M.V.; Podjava, A.; Belyakov, S. Crystal and Molecular Structure and Stability of Isoniazid Cocrystals with Selected Carboxylic Acids. *Cryst. Growth Des.* **2013**, *13*, 1082–1090. [CrossRef]
27. Swapna, B.; Maddileti, D.; Nangia, A. Cocrystals of the Tuberculosis Drug Isoniazid: Polymorphism, Isostructurality, and Stability. *Cryst. Growth Des.* **2014**, *14*, 5991–6005. [CrossRef]
28. Etter, M.C.; MacDonald, J.C.; Bernstein, J. Graph-Set Analysis of Hydrogen-Bond Patterns in Organic Crystals. *Acta Crystallogr. B* **1990**, *46*, 256–262. [CrossRef]
29. Etter, M.C. Encoding and Decoding Hydrogen-Bond Patterns of Organic Compounds. *Acc. Chem. Res.* **1990**, *23*, 120–126. [CrossRef]
30. Farrugia, L.J. *WinGX* and *ORTEP for Windows*: An Update. *J. Appl. Crystallogr.* **2012**, *45*, 849–854. [CrossRef]
31. Jaakola, V.-P.; Griffith, M.T.; Hanson, M.A.; Cherezov, V.; Chien, E.Y.T.; Lane, J.R.; Ijzerman, A.P.; Stevens, R.C. The 2.6 Angstrom Crystal Structure of a Human A2A Adenosine Receptor Bound to an Antagonist. *Science* **2008**, *322*, 1211–1217. [CrossRef]
32. Bonifácio, M.J.; Archer, M.; Rodrigues, M.L.; Matias, P.M.; Learmonth, D.A.; Carrondo, M.A.; Soares-Da-Silva, P. Kinetics and Crystal Structure of Catechol-o-Methyltransferase Complex with Co-Substrate and a Novel Inhibitor with Potential Therapeutic Application. *Mol. Pharmacol.* **2002**, *62*, 795–805. [CrossRef]
33. Binda, C.; Hubálek, F.; Li, M.; Herzig, Y.; Sterling, J.; Edmondson, D.E.; Mattevi, A. Binding of Rasagiline-Related Inhibitors to Human Monoamine Oxidases: A Kinetic and Crystallographic Analysis. *J. Med. Chem.* **2005**, *48*, 8148–8154. [CrossRef]
34. Chou, T.-H.; Epstein, M.; Michalski, K.; Fine, E.; Biggin, P.C.; Furukawa, H. Structural Insights into Binding of Therapeutic Channel Blockers in NMDA Receptors. *Nat. Struct. Mol. Biol.* **2022**, *29*, 507–518. [CrossRef]
35. Pettersen, E.F.; Goddard, T.D.; Huang, C.C.; Couch, G.S.; Greenblatt, D.M.; Meng, E.C.; Ferrin, T.E. UCSF Chimera–a Visualization System for Exploratory Research and Analysis. *J. Comput. Chem.* **2004**, *25*, 1605–1612. [CrossRef]
36. Wallace, A.C.; Laskowski, R.A.; Thornton, J.M. LIGPLOT: A Program to Generate Schematic Diagrams of Protein-Ligand Interactions. *Protein Eng. Des. Sel.* **1995**, *8*, 127–134. [CrossRef]

MDPI

Article

Studies on the Crystal Forms of Istradefylline: Structure, Solubility, and Dissolution Profile

Yiyun Wang [1,2], Youwei Xu [2,3], Zhonghui Zheng [2], Min Xue [1,*], Zihui Meng [1,*], Zhibin Xu [1], Jiarong Li [1] and Qing Lin [4]

1 School of Chemistry and Chemical Engineering, Beijing Institute of Technology, Beijing 102488, China; wangyiyun@xhzy.com (Y.W.); zbxu@bit.edu.cn (Z.X.); jrli@bit.edu.cn (J.L.)
2 Shandong Xinhua Pharmaceutical Co., Ltd., Zibo 255086, China; 3120185678@bit.edu.cn (Y.X.); zhengzhonghui@xhzy.com (Z.Z.)
3 Xiangya School of Pharmaceutical Sciences, Central South University, Changsha 410013, China
4 ReadCrystal Biotech Co., Ltd., Suzhou 215505, China; qing.lin@readcrystal.com
* Correspondence: minxue@bit.edu.cn (M.X.); mengzh@bit.edu.cn (Z.M.)

Abstract: Istradefylline as a selective adenosine A_{2A}-receptor antagonist is clinically used to treat Parkinson's disease and improve dyskinesia in its early stages. However, its crystal form, as an important factor in the efficacy of the drug, is rarely studied. Herein, three kinds of crystal forms of istradefylline prepared from ethanol (form I), methanol (form II), and acetonitrile (form III) are reported by use of a crystal engineering strategy. These three crystal forms were characterized and made into tablets for dissolution testing. Both the solubility and the dissolution rates were also determined. The dissolution rate of form I and form III is significantly higher than form II at pH 1.2 (87.1%, 58.2%, and 87.7% for form I, form II, and form III, respectively), pH 4.5 (88.1%, 58.9%, and 87.1% for form I, form II, and form III, respectively) and pH 6.8 (87.5%, 58.2%, and 86.0% for form I, form II, and form III, respectively) at 60 min. Considering the prepared solution and the proper dissolution profile, form I is anticipated to possess promising absorption for bioavailability.

Keywords: Parkinson's disease; istradefylline; solubility; crystal form; dissolution

Citation: Wang, Y.; Xu, Y.; Zheng, Z.; Xue, M.; Meng, Z.; Xu, Z.; Li, J.; Lin, Q. Studies on the Crystal Forms of Istradefylline: Structure, Solubility, and Dissolution Profile. *Crystals* **2022**, *12*, 917. https://doi.org/10.3390/cryst12070917

Academic Editors: Abel Moreno and Brahim Benyahia

Received: 15 April 2022
Accepted: 26 June 2022
Published: 28 June 2022

1. Introduction

The adenosine A_{2A}-receptor is closely related to Parkinson's disease (PD). A suitable antagonist could enhance the function of dopamine on D2 receptor neurons and result in some anti-Parkinson's effect [1–6]. Istradefylline (Figure 1) (KW-6002, (*E*)-8-(3,4-dimethoxystyryl)-1,3-diethyl-7-methyl-dihydro-1*H*-purine-2,6-dione) is the first approved adenosine A_{2A}-receptor antagonist that can improve the motor function of PD patients through its neuronal activity [7,8]. Moreover, it has also received extensive attention in pharmacology. Istradefylline was reported as a promising drug for movement disorders treatment [9]. In addition, Shin-ichi Uchida reported that istradefylline enhances the anti-parkinsonian activity of low doses of dopamine agonists [10–13].

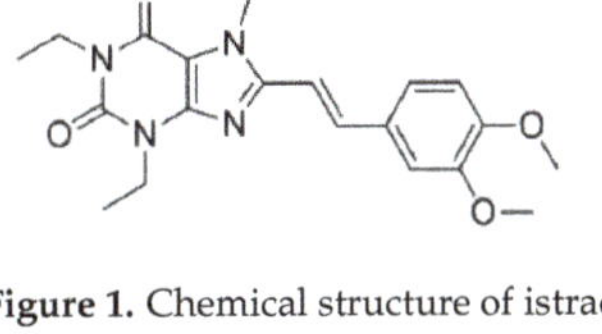

Figure 1. Chemical structure of istradefylline.

It is known that the originality of the pharmacological activity of a drug has an important influence on the effective absorption and utilization of the drug in the body.

It is remarkable that the solubility and dissolution of a drug in oral tablets significantly affects its absorption and metabolism in the body. The particle size and crystal form of a drug also affect pharmacological efficacy, due to their ability to alter the physicochemical properties of solubility, dissolution, and dosage forms. During the crystallization of a drug, different crystal structures can be formed, as the packing of molecules in space change at different temperatures, solutions, and pressures [14–17]. Generally, the appearance, melting point, dissolution, and other aspects of the same drug are significantly different in diverse crystal forms, which correspondingly affect clinical efficacy [18–22]. However, there are few reports on the solubility, crystal form, and dissolution of istradefylline.

Drug crystallization form depends on many factors, such as solvent, temperature and cooling rate, stirring speed and time, water content in solvent, and impurities in product. Based on descriptions in the literature [23–27], five crystal forms of istradefylline from ethanol/THF/isopropanol/n-propanol, methanol, acetonitrile, dichloromethane, and DMF/H_2O have been reported, while the melting points, acceleration tests, and long-term stability studies of three forms have been described; namely, the melting points of form I, form II, and form III were reported as 191.93 °C, 191.14 °C, and 191.14 °C by DSC analysis, respectively. However, the particle size, physical properties, and single crystal data of crystal forms were not reported in these patents, except for PXRD. Generally, dichloromethane and DMF were excluded in the manufacturing process due to their harmful impact on the quality of the medicine of the solution.

In this paper, we primarily discuss the dissolution rate of istradefylline in consideration of its adsorption in pharmacokinetics. In order to avoid the influence of other factors, the solubility, crystal form, particle size, and physical and chemical properties of istradefylline were also studied. The solubility and crystal form of istradefylline in seven single solvents and five mixed solvents were studied in a temperature range from 293.15 K to 333.15 K. Based on the solubility and powder diffraction data, three different crystal forms of istradefylline were obtained from ethanol (form I), methanol (form II), and acetonitrile (form III). These were consistent with the three crystal forms reported by patent NO. CN104744464A [23]. In addition, all of them were characterized by solubility, HPLC analysis, TGA and FT-IR [28,29]. In order to keep the particle size and specific surface area roughly uniform, each of the istradefylline forms was ground in a mortar for five minutes before the tablet preparation. Then, the dissolution rate of istradefylline was investigated according to the "Guidelines for determination and comparison of dissolution curves of common oral solid preparations" of Chinese pharmacopoeia. In this study, the dissolution rates of form I and form III were significantly higher than that of form II. These results have guiding significance for istradefylline tablet production.

2. Materials and Methods

2.1. Materials

Istradefylline with a purity of 99.5% was provided by Shandong Xinhua pharmaceutical Co., Ltd., Zibo, China. Acetonitrile, methanol, ethanol, ethyl acetate, n-propanol, isopropanol, and n-butanol were purchased from J.T. Baker Co., Ltd. without further purification (analytical pure), Shanghai, China. Hydrochloric acid, sodium hydroxide, potassium dihydrogen phosphate, sodium acetate trihydrate, and acetic acid were purchased from Sinopharm Group Chemical Reagent Co., Ltd. without further purification (analytical pure), Shanghai, China. Lactose (Lactose Anhydrous, NF DTHV) was provided by Kerry Inc.-Rothschild, Shanghai, China. Microcrystalline cellulose (MCC, Microcrystalline Cellulose, VIVAPUR®, PH 102) was provided by J. Rettenmaier & Sohne GmbH + Co. KG, Germany. Crospovidone (PVPP, Kollidon®, CL-F) was provided by BASF SE. Magnesium stearate (LIGAMED®, MF-2-V) was provided by Peter Greven Nederland CV. Sodium laurylsulfonate (SDS) was purchased from J&K Chemicals, Beijing, China. Purified water (18.25 $M\Omega \cdot cm^{-1}$) was obtained from a Millipore Mili-Q Plus water system. All saturated solutions prepared for HPLC detection were filtered through 0.22 μm filter membrane before usage.

2.2. HPLC analysis

The qualitative and quantitative determinations of istradefylline were performed on a Shimadzu HPLC system (Kyoto, Japan) comprising of two LC-20AT pumps, one SPD-20 UV detector, and a SIL-10A auto-sampler. The liquid chromatographic condition was optimized on an Agilent ZORBOX C18 chromatographic column (150 mm × 4.6 mm, 5 μm) with acetonitrile and water (60/40, *v*/*v*) as the stationary and mobile phase, respectively. The flow rate was confirmed as 1.0 mL·min^{-1}, while the UV-determined wavelength was 355 nm, and the sample injection volume was 20 μL.

2.3. Solubility of Istradefylline in Diverse Organic Solvents and Solvent Mixtures with Water

A certain amount of istradefylline powder was placed into a glass vial with 10 mL of acetonitrile, methanol, ethanol, ethyl acetate, n-propanol, isopropanol, n-butanol, methanol/water (30/70, *v*/*v*), ethanol/water (24/76, *v*/*v*), ethanol/water (55/45, *v*/*v*), acetonitrile/water (25/75, *v*/*v*), and acetonitrile/water (58/42, *v*/*v*), respectively. Then, the vials were incubated in a thermostat water bath for 12 h with magnetic stirring at 293.15 K, 303.15 K, 313.15 K, 323.15 K, and 333.15 K, each measured by a thermometer inside each glass vial. The temperature fluctuation of the thermostat water bath was controlled within ±0.5 K with temperature uncertainty of ±0.1 K. Then, all solutions were left to stand for a further 12 h at the corresponding temperature until the dissolution equilibrium was obtained. Then, 2 mL of supernatant from each vial was withdrawn by a syringe with a 0.22 μm filter membrane for HPLC analysis. All of the experiments were carried out three times simultaneously to obtain data averages (Table S1).

2.4. Preparation of Single Crystal

First, 1 g istradefylline was added into each of three 100 mL single-mouth flasks with 20 mL ethanol, 50 mL methanol, or 20 mL acetonitrile, respectively. Then, the mixture was stirred and heated at 78 °C, 64.5 °C and 81.0 °C, respectively, until completely dissolved, followed by being cooled down to room temperature. Stirring continued for 2 h for crystallization. Consequently, forms I, II, and III of istradefylline were obtained by filtration.

2.5. X-ray Diffraction

Single-crystal X-ray diffraction data were collected using a Bruker apex2 X-ray diffractometer equipped with a Mercury CCD detector with graphite monochromated Mo-K α radiation (λ = 0.71073 Å) at 296 K. The structures were solved by direct methods and refined by full-matrix least-squares on F2 values (SHELXL-97). Non-hydrogen atoms were refined anisotropically. Hydrogen atoms were fixed at calculated positions and refined using a riding mode. Powder X-ray diffraction (PXRD) patterns of samples were collected on a Bruker D8 Focus X-ray diffractometer with Cu Kα radiation (λ = 1.54 Å) at a scanning rate of 0.02° s^{-1} from 5° to 50° in 2θ.

2.6. Differential Scanning Calorimetry (DSC) and Thermogravimetric Analysis (TGA)

The DSC and TGA was determined by TGA/DSC1/1100LF(Mettler Toledo, Switzerland). The temperature range was 25~1100 °C; temperature accuracy was ±0.3 °C; calorimetric accuracy was ±1%; balance sensitivity was 0.1 μg; heating rate was 0.1~100 K/min.

The experiment was performed under N_2 atmosphere at 1 atm with a heating rate of 10 °C/min in a temperature range of 30~400 °C.

2.7. Fourier-Transform Infrared Spectral Analysis (FT-IR)

FT-IR analysis was collected in a range of 3600–1600 cm^{-1} using KBr pellets and a Thermo iD7 ATR infrared spectrometer (Thermo Fisher Technology (China) Co., Ltd., Shanghai, China).

2.8. Particle Size and Specific Surface Area Analysis (BET)

The particle sizes of forms I/II/III were determined by a Malvern 2000 laser particle size analyzer (Malvern, England). The specific surface areas (N_2 adsorption) of forms I/II/III were detected by the specific surface-area analyzer BK200B (Beijing Jingwei Gaobo Science and Technology Co., Ltd., Beijing, China.).

The samples were mixed thoroughly and evenly (loose clumps were gently pressed with a spoon to completely disperse) and flatly laid on the sample table of the Scirocco 2000 dry sampler. Vibration injection speed was 30~80%; relaxation-dispersed air pressure was 2.5 bar; cost of shading was 1~5%; measuring time was 10 s; background time was 10 s.

Specific surface-area analysis was performed under N_2 adsorption with adsorption temperature of 77.35 K on a BK200B using the static capacity method, while the temperature was controlled at 40 °C for 360 min.

2.9. Dissolution Study

The instruments used for tablets included a circulating water vacuum pump (SHB-III, Zhengzhou Great Wall Science, Industry and Trade Co., Ltd., Zhengzhou, China), an electrothermal blast drying box (GZX-9240MBE, Shanghai Boxun Industrial Co., Ltd., Shanghai, China), an electronic balance (PB3002-S, METTLER TOLEDO), a constant-temperature magnetic stirrer (DF-101S, Zhengzhou Great Wall Industry and Trade Co., Ltd.), an ultraviolet-absorption spectrophotometer (UV1800, Shimadzu, Kyoto, Japan), a dissolution tester (SNTR-8400AT, Shimadzu), a single-stamping-sheet machine (YP-1, HangZhou XuZhong Food Machinery Co., Ltd.), and a high-efficiency coating machine (JCB/K-3/5/10, Wenzhou Jianpai Pharmaceutical Machinery Co., Ltd., Wenzhou, China, nozzle diameter of 1 mm).

In order to prepare the istradefylline tablets, a prescribed amount of istradefylline was weighed and ground in a mortar for 5 min. Lactose, microcrystalline cellulose PH102, and PVPP were weighed and mixed with istradefylline by hand for 3 min. Then, magnesium stearate (MS) was weighed and added into the above mixture and mixed again by hand for 3 min. A single-stamping-sheet machine (mold circular concave Φ7.1 mm, tablet weight 140 mg, tablet hardness −5 kp) was used for tableting. The coating liquid prepared by Opadry 03K19229 (solid content: 8%) was coated on a high-efficiency coating machine. The inlet air temperature was 70 °C, the atomization gas pressure was 0.3 MPa, the rotation speed was 8 rpm, and the spraying speed was 7 rpm. Coating-weight gain was controlled at about 3%. The batch size was 1000 pieces. The tablet speed was 1000 tablets/h, and the coating batch was 800 tablets/batch. The spray speed was 7.0 g/min. The prescription ingredients are shown in Table 1 in detail.

Table 1. Prescription ingredients list.

Process	Material Name	Function	Batch Size/g	Proportion/%
Tablet	Istradefylline	Drug	20.00	14.29
	MCC PH102	Fillers	67.50	48.21
	Lactose	Diluents	42.00	30.00
	PVPP	Disintegrant	9.80	7.00
	MS	Lubricants	0.70	0.50
	Total		140.0	100.0
Coating	Opadry	Materials	5.2	3%
Single piece of content	Purified water	Solvent	59.8	Final removal
Actual use	Opadry	Materials	54.4	3%
(1.2 times preparation)	Purified water	Solvent	626.0	Final removal

According to the dissolution and release determination method (Chinese Pharmacopoeia 2020 Edition, general rule of the fourth part 0931, second method), six tablets of istradefylline were put into a beaker filled with 900 mL buffer solution of pH 1.2, pH 4.5, and pH 6.8, while the rotation speed was fixed at 75 r/min. Then, 10 mL samples were taken out at 5 min, 10 min, 15 min, 30 min, 45 min, and 60 min and filtrated with 0.22 μm filter membrane. Then, 2 mL filtrate was diluted to 10 mL with a diluent (acetonitrile-water (50:50)), and the test sample was obtained. Next, 25 mg istradefylline was precisely weighed and dissolved into 25 mL acetonitrile in a volumetric flask. Then, 1 mL istradefylline solution was diluted to 5 mL with a diluent (acetonitrile-water (50:50)), and the solution was mixed. Next, 2 mL solution was precisely measured and transferred into a 10 mL volumetric flask. Then, 2 mL dissolution medium was added, and the solution was diluted to scale with a diluent (acetonitrile-water (50:50)), upon which the reference solution was obtained.

The same amount of diluent (acetonitrile-water (50:50)) was added to the two cuvettes, and then they were placed in channel 1 and channel 2 of UV-Vis spectrophotometer, respectively. After the instrument was zeroed, the reference solution and sample solution were placed in channel 2 and measured at a wavelength of 362 nm. Each group of samples was repeatedly tested six times, and the RSD of all samples at each time point was less than 10%, which proved that each sample had good uniformity.

Computational Formula:

$$\text{Dissolution} = \frac{A_{test} \times C_{reference} \times 5 \times 900}{A_{reference} \times 20} \times 100\%$$

$$\text{Cumulative dissolution} = A_n + \frac{(A_n - 1 + \ldots\ldots + A_1) \times 10}{900}$$

where A_{test} is UV absorbance of sample, $A_{reference}$ is UV absorbance of reference substance, $C_{reference}$ is concentration of reference substance.

3. Results and Discussion

3.1. Solubility of Istradefylline

A perfect chromatogram of istradefylline as a symmetrical sharp peak was obtained, as shown in Figure 2. The relationship between the chromatogram peak area and concentration expressed as calibration curve is graphically displayed in Figure 2. The linear fitting equation was Y = 4.42X + 7.06 with a concentration range of 0.001 mg·mL^{-1} to 0.1 mg·mL^{-1}, while the linear dependence was 0.9999. The linearity was used to calculate the istradefylline concentration in the supernatant of each vial in the experiment by HPLC detection.

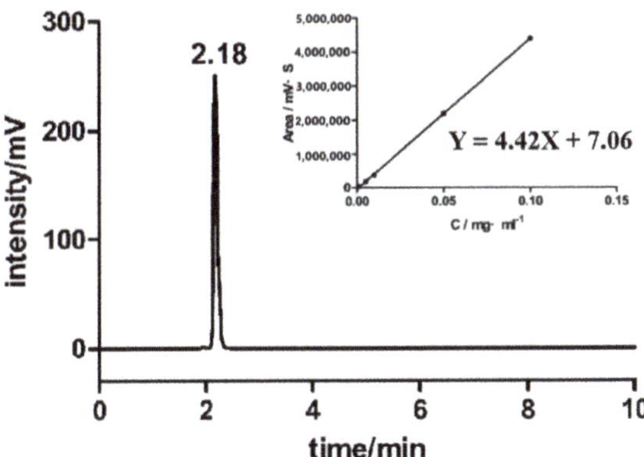

Figure 2. HPLC chromatogram of istradefylline. Insets show the linear relationship between the chromatogram peak area (Y) and the concentration (X) of istradefylline in acetonitrile.

In this study, the solubility data of istradefylline in common organic solvents in the range of 293.15 K to 333.15 K were determined by an established HPLC method with milligram-grade usage. The solubility of istradefylline was expressed by mole fraction of the solute in the solution. The mass of the solute in the sample solution can be calculated according to Equation (1), while the concentration of istradefylline in saturated solution was estimated by the liquid chromatographic method according to the calibration curve,

$$m = c \cdot v \quad (1)$$

where m is the mass of istradefylline in saturated solution, c is the corresponding concentration, and v is the volume after diluted. The mole fraction of the solute can be readily calculated as follows:

$$x = \frac{m_1/M_1}{m_1/M_1 + (m_0 - m_1)M_2} \quad (2)$$

where x is the mole fraction of the solute istradefylline, m_1 is the mass of the solute calculated by Equation (1), M_1 is the molecular weight of solute, m_0 is the mass of the solution, and M_2 is the molecular weight of solvent. The precise solubility of this compound in seven single-solvents and five mixed-solvents in the range of 293.15 K to 333.15 K is recorded in Table S1. Furthermore, the temperature influence on the solubility of istradefylline was also studied. Solubility increased at an exponential rate with rising temperature in all solvents, as shown in Figure 3. Generally, the solubility of chemicals is an endothermic process, so increasing the temperature is beneficial for increasing the solubility of drugs. Besides the solubility data, which were useful in the quality control and process improvement that followed, three kinds of crystalline forms were also determined and classified in these solvents by powder X-ray diffraction method.

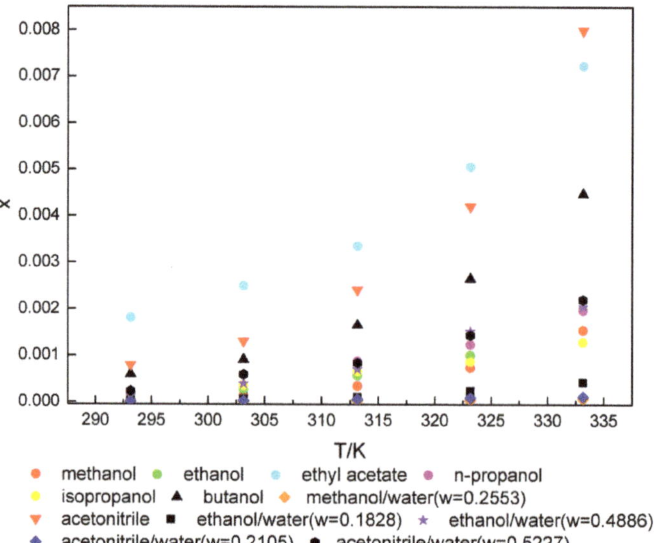

Figure 3. Temperature dependence of mole fraction of istradefylline in several solvents.

It is known that the solubility of istradefylline is very low in aqueous media in the pH range from 1.0 to 12.0, so the solubility of istradefylline in aqueous solutions of different pH was tested. The test results were shown in Table 2. From the test results, it can be seen that the solubility of istradefylline decreased when the pH increased, since istradefylline is a weakly alkaline drug that has greater solubility in acidic solutions.

Table 2. Solubility of istradefylline in different pH at 293.15 K.

pH	Solubility of Istradefylline (μg/mL)
1	0.41
2	0.39
3	0.32
4	0.31
7	0.27
8	0.18
10	0.11
12	0.10

3.2. Powder X-ray Diffraction

The istradefylline crystalline solids from the saturated solutions were characterized by X-ray powder diffraction (Figure 4). Three forms can be clearly distinguished from the significant differences present among their diffraction patterns. Form I could be obtained in a wide variety of solvent systems, including ethyl acetate, n-propanol, isopropanol, n-butanol, ethanol, ethanol/water (w = 0.1828), and ethanol/water (w = 0.4886). Its powder diffraction pattern is characterized by peaks at 2θ = 6.98°, 11.02°, 13.98°, 15.68°. Form II crystallized in methanol and methanol/water (w = 0.2553). Its characteristic diffraction peaks can be found at 2θ = 8.68°, 11.86° and 12.12°. Form III, obtained in acetonitrile, acetonitrile/water (w = 0.2105), and acetonitrile/water (w = 0.5127), shows characteristic diffraction peaks at 2θ = 9.74°, 10.24°, 12.38° and 25.07°. These results were consistent with the three forms disclosed by patent number CN104744464A.

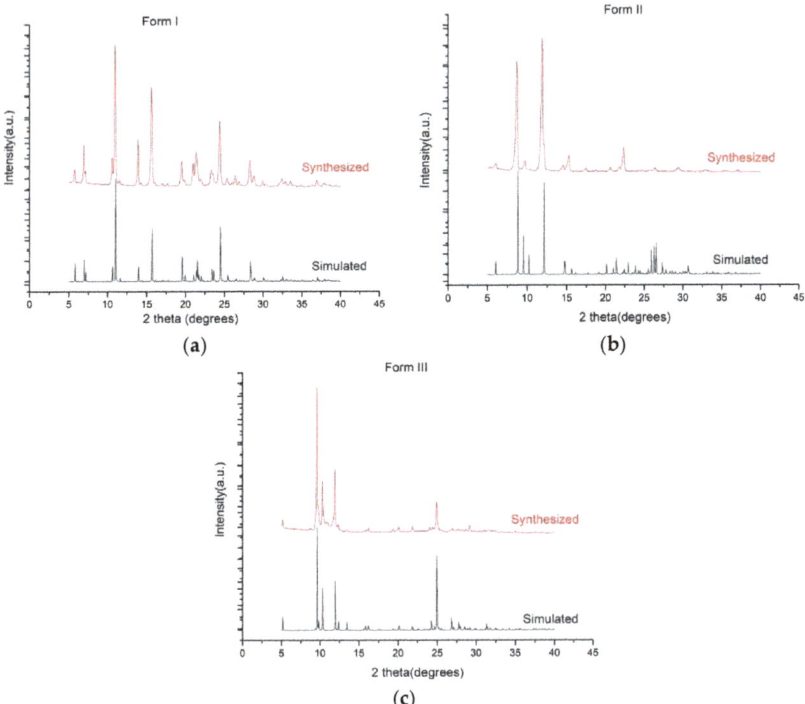

Figure 4. Comparison of powder X-ray diffraction (after grinding) and crystallographic data of istradefylline. (**a**) Form I made in ethanol; (**b**) Form II made in methanol; (**c**) Form III made in acetonitrile.

3.3. *Single-Crystal X-ray Diffraction*

For further study, three forms were obtained using ethanol, methanol, and acetonitrile as the crystallization solvents, respectively (CCDC number: 2043873-2043875). Their crystal structure and crystallography data are shown in Figures S1–S3 and Table 3.

Table 3. Crystallographic data and structure refinement parameters.

Compounds	Form I	Form II	Form III
Chemical formula	$C_{20}H_{24}N_4O_4$	$C_{20}H_{26}N_4O_5$	$C_{22}H_{29}N_5O_5$
Formula weight	384.43	402.45	443.50
Crystal system	monoclinic	monoclinic	monoclinic
Space group	*P*21	*P*21/*c*	*P*21/*m*
a/Å	13.6762(17)	4.5436(5)	9.430(9)
b/Å	4.7483(7)	23.776(2)	7.129(7)
c/Å	16.464(2)	18.6282(17)	17.587(18)
α/°	90	90	90
β/°	112.39(4)	95.065(7)	103.514(10)
γ/°	90	90	90
vol/Å3	988.5(2)	2004.6(3)	1149.5(19)
Z	2	4	2
ρcalcg/cm^3	1.292	1.334	1.281
Gof	1.017	1.036	1.052
R	R1= 0.0552, wR2 = 0.1107	R1 = 0.0888, wR2 = 0.2892	R1 = 0.083, wR2 = 0.27

According to the crystallographic data and structure refinement parameters, three forms can be effectively distinguished. From the single-crystal structure, it can be seen that form I was pure crystal, form II was monohydrate crystal, and form III was monohydrate of acetonitrile solvent complex. The three forms were all monoclinic systems with different space groups, with form I, II and III exhibiting *P*21, *P*21/*c* and *P*21/*m*, respectively. The molecular packing and intermolecular interactions of the three forms were different (Figure 5). Form I has strong intermolecular hydrogen bonds between oxygen atoms and the nearby hydrogen on the other side of the molecule (C=O . . . H, 2.37 Å and 3.16 in form I), without π-π stacking. Form II and form III have hydrogen bonds between oxygen atoms and the nearby hydrogens of water molecules (O5=O2 . . . H, 2.878 Å in form II and O5A=O3 . . . H, 2.955 Å in form III). Although abundant hydrogen bonds were constructed through the interactions between water molecules and the crystal of form II and form III, their bond energies of about 15~30 kJ·mol^{-1} were much lower than those of general chemical bonds. This suggests that these hydrogen bonds are fragile, and the water molecules are transferred during the drying process of istradefylline. Otherwise, form II was a head-to-head π-π interaction, and form III was a head-to-tail π-π interaction. The π-π stacking interaction of form II, at 3.55 Å, was stronger than that of form III, at 3.54 Å, between two molecules. It is well-known that π-π stacking is detrimental to the solubility of compounds. Different π-π stacking forms also have an effect on solubility, which may lead to the weak solubility and dissolution of the crystalline form II of istradefylline.

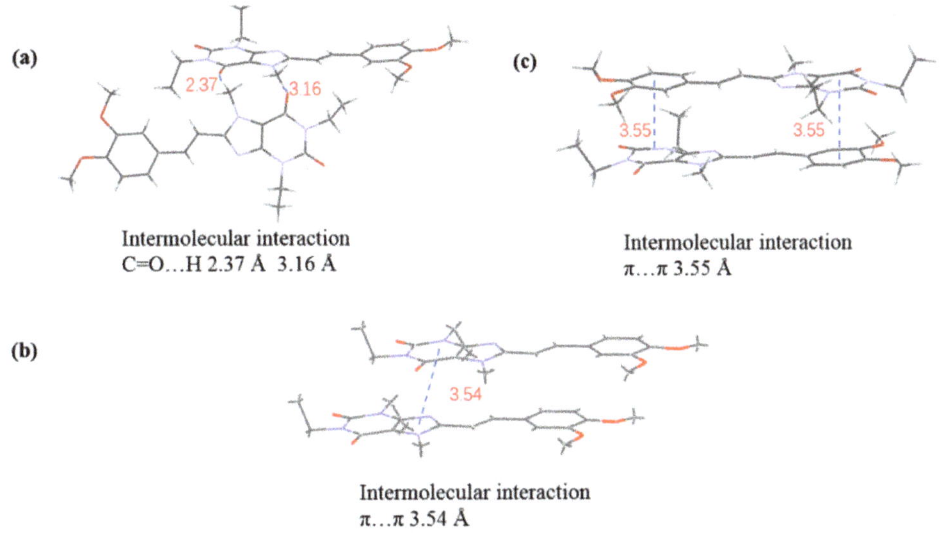

(a)

Intermolecular interaction
C=O...H 2.37 Å 3.16 Å

(c)

Intermolecular interaction
π...π 3.55 Å

(b)

Intermolecular interaction
π...π 3.54 Å

Figure 5. Single-crystal molecular structure of istradefylline in (**a**) ethanol (form I), (**b**) methanol (form II) and (**c**) acetonitrile (form III), respectively, and π-π stacking and hydrogen-bond lengths.

3.4. TGA

Figure S5 shows the TGA curves for form I, form II, and form III at a heating rate of 10 °C/min under N_2 atmosphere. As shown in Figure S5a, the weight loss of form I was 99.98% from 191.66 to 198.08 °C, corresponding to the degradation of istradefylline molecules. In Figure S5b, the weight loss of form II was in two stages: the first weight loss of 3.23% from 27.56 °C to 52.45 °C corresponds to the release of one H_2O molecule (calc. 4.48%), while the second weight loss occurred at or above 377.72 °C, corresponding to the degradation of istradefylline. In Figure S5c, the thermal decomposition process of cocrystal III was in two stages: the first weight loss of 15.62% from 91.16 °C to 121.02 °C corresponds to the release of one H_2O molecule and one CH_3CN molecule (calc. 13.32%), while the second weight loss occurred at or above 258.93 °C, corresponding to the degradation of istradefylline.

3.5. FT-IR Analysis

In Figure S6a, the absorbance peaks at 2966 cm^{-1}, 2935 cm^{-1}, and 2832 cm^{-1} are ascribed to the presence of the methyl or methylene group. In Figure S6b, the absorbance peaks at 3482 cm^{-1}, 2977 cm^{-1}, 2935 cm^{-1}, and 2841 cm^{-1} are ascribed to the presence of the hydroxy, methyl, or methylene group. In Figure S6c, the absorbance peaks at 2979 cm^{-1}, 2932 cm^{-1}, 2839 cm^{-1}, and 2217 cm^{-1} are ascribed to the presence of the methyl or methylene group, the absorbance peaks at 3465 cm^{-1}, 3380 cm^{-1} and 3028 cm^{-1} indicate the presence of water molecules, and the absorbance peak at 2217 cm^{-1} indicates the presence of acetonitrile molecules.

3.6. Particle Size and BET Analysis

The particle size of form I, II and III was determined with a Malvern 2000 laser particle-size analyzer, and the median particle sizes were 5.6 μm, 5.2 μm, and 7.1 μm, respectively. The specific surface areas of the three forms were 5.08 m^2/g, 5.51 m^2/g and 5.18 m^2/g, respectively. The specific surface areas of single points were 0.20000 at P/Po is 4.67 m^2/g, 4.74 m^2/g, and 4.72 m^2/g, respectively. As can be seen from the above data, the surface area was basically the same.

3.7. Dissolution Curve Test

The dissolution of istradefylline in three crystal forms was studied in a buffer solution in which pH value was fixed at 1.2, 4.5, and 6.8, corresponding to pH of digestive solutions such as gastric juice and intestinal juice. All experiments were repeated six times, as shown in Figure 6. The dissolution rates of the three crystal forms at pH 1.2 at 5 min were 72.2%, 28.3%, and 74.7%, respectively. At 60 min, the dissolution rates of form I, II, and III reached 87.1%, 58.2%, and 87.7% respectively. The dissolution rates of the three crystal forms at pH 4.5 at 5 min were 68.3%, 29.4%, and 73.2%, respectively. At 60 min, the dissolution of form I reached 88.1%, the dissolution of form II reached 58.9%, and the dissolution of form III reached 87.1%. The dissolution rates of the three crystal forms at pH 6.8 at 5 min were 69.2%, 30.3%, and 70.7%, respectively. At 60 min, the dissolution of form I reached 87.5%, the dissolution of form II reached 58.2%, and the dissolution of form III reached 86.0%. From the dissolution profile of the three crystal forms of istradefylline, it can be inferred that the dissolution of istradefylline form I, refined in ethanol, and form III, refined in acetonitrile, show good dissolution performance, while that of form II, refined in methanol, was significantly lower. Through single-crystal studies, it was found that form II has π-π stacking, and the π-π stacking interaction of form II is stronger than that of form III between two molecules. The π-π stacking leads to weak solubility and dissolution in istradefylline. These results indicate that the solvent has a direct effect on the dissolution.

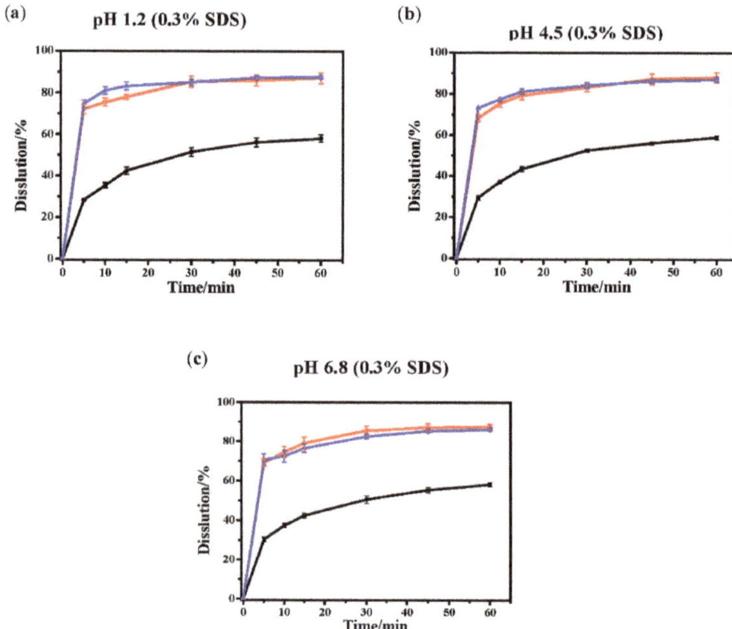

Figure 6. Dissolution curve of three crystal forms of istradefylline.

4. Conclusions

In this paper, the solubility of istradefylline in 12 kinds of solvents such as ethanol, methanol, and acetonitrile was studied, and three kinds of crystal forms were sequentially obtained in these solvents and proved by X-ray powder diffraction. Their single-crystal diffraction structure and data were also confirmed by single-crystal diffraction. Furthermore, the dissolution test was performed with tablets prepared from the three crystal forms of istradefylline, and it was found that the dissolution rates of the three crystal forms were different. Compared with form II, the dissolution rates of form I and form III were superior. Additionally, the solubility and X-ray powder diffraction data, as well as the dissolution rates, indicated that the packing of molecules and crystal forms have notable influence on solubility and dissolution. This study offers additional insight into optimizing the crystallization process of istradefylline and improving absorption in pharmacokinetics.

5. Patents

Preparation method and application of istradefylline crystal, CN113024558A, 2021-06-25.

Supplementary Materials: The following supporting information can be downloaded at: https://www.mdpi.com/article/10.3390/cryst12070917/s1. Table S1. Solubility of Istradefylline in Twelve Different Solvents from 293.15 K to 333.15 K. Table S2. Bond Lengths for Istradefylline. Table S3. Stability data of Istradefylline (Form I of istradefylline). Figures S1–S3. ORTEP view with labeling scheme for Istradefylline. Figure S4. The surface area of three crystal forms. Figure S5. (a) The TGA curves of From I, (b) The TGA curves of From II and (c) The TGA curves of From III. Figure S6. The IR of From I, The IR of From II and The IR of From III.

Author Contributions: Conceptualization, Y.W. and M.X.; methodology, Y.W.; software, M.X.; validation, M.X.; formal analysis, Z.Z., Z.M., Z.X., J.L. and Q.L.; investigation, Y.W.; resources, Y.W.; data curation, Y.X.; writing—original draft preparation, Y.W. and M.X.; writing—review and editing, Y.W. and M.X. All authors have read and agreed to the published version of the manuscript.

Funding: This research received no external funding.

Institutional Review Board Statement: Not applicable.

Informed Consent Statement: Not applicable.

Data Availability Statement: Not applicable.

Acknowledgments: The authors wish to thank the Suzhou Jingyun Pharmaceutical Technology Co., Ltd., Suzhou ReadCrystal Biotechnology Co., Ltd. and the instrumental analysis center of Beijing Institute of Technology for their powder X-ray diffraction analysis, single crystal diffraction and so on.

Conflicts of Interest: The authors declare that they have no known competing financial interest or personal relationships that could have appeared to influence the work reported in this paper.

References

1. Zheng, J.Y.; Zhang, X.H.; Zhen, X.C. Development of Adenosine A_{2A} Receptor Antagonists for the Treatment of Parkinson's Disease: A Recent Update and Challenge. *ACS Chem. Neurosci.* **2019**, *10*, 783–791. [CrossRef]
2. Shook, B.C.; Jackson, P.F. Adenosine A_{2A} Receptor Antagonists and Parkinson's Disease. *ACS Chem. Neurosci.* **2011**, *2*, 555–567. [CrossRef]
3. Hauser, R.A.; Shulman, L.M.; Trugman, J.M.; Roberts, J.W.; Sussman, N.M. Study of istradefylline in patients with Parkinson's disease on levodopa with motor fluctuations. *Mov. Disord.* **2010**, *23*, 2177–2185. [CrossRef]
4. Tian, S.; Wang, X.; Li, L.L.; Zhang, X.H.; Li, Y.Y.; Zhu, F.; Hou, T.J.; Zhen, X.C. Discovery of Novel and Selective Adenosine A2A Receptor Antagonists for Treating Parkinson's Disease through Comparative Structure-Based Virtual Screening. *J. Chem. Inf. Model.* **2017**, *57*, 1474–1487. [CrossRef]
5. Lu, J.; Cui, J.; Li, X.H.; Wang, X.; Zhou, Y.; Yang, W.J.; Chen, M.; Zhao, J.; Pei, G. An Anti-Parkinson's Disease Drug via Targeting Adenosine A_{2A} Receptor Enhances Amyloid-β Generation and γ-Secretase Activity. *PLoS ONE* **2016**, *11*, e0166415. [CrossRef]
6. Uchida, S.I.; Soshiroda, K.; Okita, E.; Uchida, M.K.; Mori, A.; Jenner, P.; Kanda, T. The adenosine A_2A receptor antagonist, istradefylline enhances the anti-parkinsonian activity of low doses of dopamine agonists in MPTP-treated common marmosets. *Eur. J. Pharmacol.* **2015**, *747*, 160–165. [CrossRef]
7. Ko, W.K.D.; Sandrine, M.C.; Li, Q.; Yang, J.Z.; Steve, M.; Elsa, Y.P.; Erwan, B. An evaluation of istradefylline treatment on Parkinsonian motor and cognitive deficits in 1-methyl-4-phenyl-1,2,3,6-tetrahydropyridine (MPTP)-treated macaque models. *Neuropharmacology.* **2016**, *110*, 48–58. [CrossRef]
8. Matsuura, K.; Kajikawa, H.; Tabei, K.I.; Satoh, M.; Kida, H.; Nakamura, N.; Tomimoto, H. The effectiveness of istradefylline for the treatment of gait deficits and sleepiness in patients with Parkinson's Disease. *Neurosci. Lett.* **2018**, *662*, 158–161. [CrossRef]
9. Kataoka, H.; Sugie, K. Does istradefylline really have a dystonic mechanism? *J. Neurol. Sci.* **2018**, *388*, 233–234. [CrossRef]
10. Fumio, S.; Junichi, S.; Nobuaki, K.; Joji, N.; Shizuo, S.; Shunji, I.; Hiromi, N. Therapeutic Agents for Parkinson's Disease. EP0590919, 29 December 1999.
11. Li, J.; Han, Y.F.; Li, Y.X.; Chen, S.W.; Gong, P. Improved synthesis process of adenosine A_2A receptor inhibitor istradefylline. *Chin. J. Med. Chem.* **2018**, *2*, 220–224.

12. Uchida, S.I.; Soshiroda, K.; Okita, E.; Uchida, M.K.; Mori, A.; Jenner, P.; Kanda, T. The adenosine A_2A receptor antagonist, istradefylline enhances anti-parkinsonian activity induced by combined treatment with low doses of L-DOPA and dopamine agonists in MPTP-treated common marmosets. *Eur. J. Pharmacol.* **2015**, *766*, 25–30. [CrossRef] [PubMed]
13. Uchida, S.I.; Tashiro, T.; Kawai-Uchida, M.; Mori, A.; Jenner, P.; Kanda, T. The adenosine A_{2A}-receptor antagonist istradefylline enhances the motor response of L-DOPA without worsening dyskinesia in MPTP-treated common marmosets. *J. Pharmacol. Sci.* **2014**, *124*, 480–485. [CrossRef] [PubMed]
14. Huang, T.H.; Lu, D.Q.; Ling, X.Q.; Wang, X.X.; Liu, T.Q.; Shen, F.F.; He, K.F. Thermodynamic models for determination of the solid-liquid equilibrium of Istradefylline in ethyl acetate plus (isopropanol, tetrahydrofuran, acetone) binary solvent mixtures. *J. Chem. Thermodyn.* **2017**, *111*, 31–40. [CrossRef]
15. Ge, Y.H.; Li, T.T.; Cheng, J.J. Crystal Type I of Azilsartan Polymorphs: Preparation and Analysis. *J. Cryst. Process. Technol.* **2016**, *6*, 1–10. [CrossRef]
16. Yadav, M.R.; Shaikh, A.R.; Ganesan, V.; Giridhar, R.; Chadha, R. Studies on the crystal forms of pefloxacin: Preparation, characterization and dissolution profile. *J. Pharm. Sci.* **2008**, *97*, 2637–2648. [CrossRef]
17. Ganesan, V. Studies on the crystal forms of moxifloxacin: Preparation, characterization and dissolution profile. *Pharm. Anal. Acta* **2013**, *4*, 135.
18. Zhang, X.M.; Sun, F.X.; Zhang, T.T.; Jia, J.T.; Su, H.M.; Wang, C.H.; Zhu, G.S. Three pharmaceuticals cocrystals of adefovir: Syntheses, structures and dissolution study. *J. Mol. Struct.* **2015**, *1100*, 395–400. [CrossRef]
19. Llinas, A.; Barbas, R.; Font-Bardia, M.; Quayle, M.J.; Velaga, S.; Prohens, R. Two New Polymorphic Cocrystals of Zafirlukast: Preparation, Crystal Structure, and Stability Relations. *Cryst. Growth Des.* **2015**, *15*, 4162–4169. [CrossRef]
20. Hua, D.Y.; Chen, G.L.; Shen, W.J. Determination of two crystal forms of famotidine by differential scanning calorimetry. *J. Chin. Med. Ind.* **1991**, *22*, 78–79.
21. Wang, J.; Zhang, R.H.; Sun, S.Y. Study on the polycrystalline form of nimodipine. *Acta pharm. Sin.* **1995**, *30*, 443–448.
22. Jiao, L.T.; Zhang, L.; Yang, D.Z.; Yang, S.Y.; Du, G.H.; Lv, Y. Raman spectroscopic analysis and dissolution tests of nimodipine crystal forms. *Her. Med.* **2017**, *36*, 1175–1179.
23. Bao, J.Y.; Huang, H.; Yu, D.J.; Wei, W.; Jiang, Y.W.; Zhang, X.Q. Polymorphs of Istradefylline. CN104744464A, 1 July 2015.
24. Dong, D.D. The invention relates to a method for preparing Istradefylline crystal form III by ball milling. CN108117554A, 5 June 2018.
25. Wang, C.H. A new crystal form of Istradefylline and its preparation method. CN105884776A, 24 August 2016.
26. Gong, D.H.; Wang, J.; Gai, J.H.; Yang, J.; Sun, W.J.; Yang, M.; Yang, C.Q.; Ma, Y.X. A new crystal form of Istradefylline and its preparation method. CN106279169A, 4 January 2017.
27. Dong, D.D. A preparation method of Istradefylline crystal form II for treating Parkinson's disease. CN108101907A, 1 June 2018.
28. Bourne, S.A.; Villiers, M.D.; Crider, A.M.; Caira, M.R. Polymorphism of the antitubercular Isoxyl. *Cryst Growth Des.* **2011**, *11*, 4950–4957.
29. Cui, K.J.; Yang, Y.M.; Meng, Z.H.; Xu, G.R.; Xu, Z.B. Solubility of tetranitrodimerglycoluril (TNDGU) in different solvents at temperatures between 293.15 K and 313.15 K. *J. Chem. Eng. Data* **2014**, *59*, 2620–2622. [CrossRef]

MDPI

Article

The Effect of DNA from Escherichia Coli at High and Low CO_2 Concentrations on the Shape and Form of Crystal-line Silica-Carbonates of Barium (II)

Cesia D. Pérez-Aguilar [1], Selene R. Islas [2], Abel Moreno [3,*] and Mayra Cuéllar-Cruz [1,*]

1 Departamento de Biología, División de Ciencias Naturales y Exactas, Campus Guanajuato, Universidad de Guanajuato, Noria Alta S/N, Col. Noria Alta, C.P., Guanajuato 36050, Mexico
2 Instituto de Ciencias Aplicadas y Tecnología, Universidad Nacional Autónoma de México, Circuito Exterior S/N, Ciudad Universitaria, Mexico City 04510, Mexico
3 Instituto de Química, Universidad Nacional Autónoma de México, Av. Universidad 3000, Col. Ciudad Universitaria, Ciudad de México 04510, Mexico
* Correspondence: carcamo@unam.mx (A.M.); mcuellar@ugto.mx (M.C.-C.)

Abstract: The synthesis of nucleic acids in the Precambrian era marked the start of life, with DNA being the molecule in which the genetic information has been conserved ever since. After studying the DNA of different organisms for several decades, we now know that cell size and cellular differentiation are influenced by DNA concentration and environmental conditions. However, we still need to find out the minimum required concentration of DNA in the pioneer cell to control the resulting morphology. In order to do this, the present research aims to evaluate the influence of the DNA concentration on the morphology adopted by biomorphs (barium silica-carbonates) under two synthesis conditions: one emulating the Precambrian era and one emulating the present era. The morphology of the synthetized biomorphs was assessed through scanning electron microscopy (SEM). The chemical composition and the crystalline structure were determined through Raman and IR spectroscopy. Our results showed that DNA, even at relatively low levels, affects the morphology of the biomorph structure. They also indicated that, even at the low DNA concentration prevailing during the synthesis of the first DNA biomolecules existing in the primitive era, these biomolecules influenced the morphology of the inorganic structure that lodged it. On the other hand, this also allows us to infer that, once the DNA was synthetized in the Precambrian era, it was definitely responsible for generating, conserving, and directing the morphology of all organisms up to the present day.

Keywords: biomorph synthesis; effects of DNA concentration on biomorphs; kerogen; Precambrian era

Citation: Pérez-Aguilar, C.D.; Islas, S.R.; Moreno, A.; Cuéllar-Cruz, M. The Effect of DNA from Escherichia Coli at High and Low CO_2 Concentrations on the Shape and Form of Crystal-line Silica-Carbonates of Barium (II). *Crystals* **2022**, *12*, 1147. https://doi.org/10.3390/cryst12081147

Academic Editor: Carlos Rodriguez-Navarro

Received: 18 July 2022
Accepted: 13 August 2022
Published: 15 August 2022

Publisher's Note: MDPI stays neutral with regard to jurisdictional claims in published maps and institutional affiliations.

1. Introduction

The prebiotic synthesis of nucleic acids from the polymerization of pyrimidine nucleotides was performed in primitive Earth due to the apparently existing conditions for the chemical synthesis of biological polymers and for the start of life [1,2]. Ribonucleic acid (RNA) has been proposed by some authors as the first biomolecule to be formed in a primigenial phase due to its catalytic activity [3–5]. It has also been proposed that, once a certain number of RNA molecules were available, a ribosome started to catalyze the formation of polypeptides; it is currently accepted that the ribosome corresponds to the large subunit of ribosomal RNA [3–7]. As for deoxyribonucleic acid (DNA), some authors propose that it must have been formed at a later stage than RNA. However, it has been recently described that both RNA and DNA must have been synthetized at the same time in the primigenial era; otherwise, it would not have been able to function had there been a sole RNA or DNA world in a prebiotic context [7–13]. DNA has, therefore, a selective advantage as it has, since then, become the molecule in charge of storing the genetic information. However, DNA also presents another characteristic, whereby its cellular concentration can be related to the cell size remaining stable even in nonfavorable nutritional and environmental conditions [14]. This characteristic has been studied in *Bacterium lactis aerogenes, Escherichia coli, Salmonella* Typhimurium, *Citrobacter freundii, Serratia marcescens, Bacillus subtilis, Erwinia carotovora, Micrococcus anhaemolyticus, Pseudomonas aeruginosa, Lactobacillus bulgaricus, Saccharomyces cerevisiae, Tetrahymena pyriformis* GL, and marine organisms, among others [14–18]. The latter is relevant because maintaining life in an organism seems to lie in its ability to maintain the DNA concentration. In this sense, in several vertebrate species, the nuclei of somatic cells contain fixed amounts of DNA [15,19]. In the microorganism *Bacterium lactis aerogenes*, it was found that the DNA is a constant constituent of the bacterial cell [15]. In other bacteria, such as *Escherichia coli*, the concentration of proteins increases in proportion to the DNA, without being affected by the growth rate. In this way, the protein/DNA relationship is independent of the growth rate [16]. Both RNA and DNA are molecules that have played a leading role in the chemical origin of life since the primigenial era of Earth. However, we still need to know the minimum required DNA to direct the morphology of the protocell. It is now known that a higher concentration of this nucleic acid in the cell indicates a more complex species. However, despite this information, we still need to know the minimal DNA concentration required to influence the morphology and characteristics of the pioneer cell. In this sense, Wächtershäuser (2006) [20] proposed that the primitive cell was formed by both an inorganic or mineral part and an organic one. Our research team has emulated this cell in a simple way by using biomorphs as study models. Calcium, barium, or strontium silica-carbonate biomorphs are self-assembled crystalline nano- or micromaterials that usually display a variety of biomimetic morphologies. These biomorphs show characteristic curvatures, which are far away from the restrictions of the classic crystallographic symmetry [21–30]. Recently, our research team showed that the biomorphs are not only interesting from the point of view of morphology but also because they could have been the first mineral structure in which the first biomolecules became isolated from the outside environment, aligned, polymerized, and conserved to give origin to the primigenial cell [29]. We also showed that DNA influences predominantly the morphology adopted by biomorphs [22–24,29]. However, the question about the minimal concentration required in pioneer cells for DNA to control the morphology to be adopted still remains. To answer this question, this work aims to evaluate, for the first time, the influence of DNA concentration on the morphology adopted by the biomorphs in two different synthesis conditions, one emulating the conditions of the Precambrian era and the other emulating those of the present one. The morphology of the synthetized biomorphs was assessed through scanning electron microscopy (SEM). The chemical composition and the crystalline structure were determined through Raman and IR spectroscopy. Our results showed that DNA, even at relatively low levels, affected the morphology of the biomorphs structure. They also indicated that, even at the low DNA concentration prevailing during the synthesis of the first DNA biomolecules existing in the primitive era, these biomolecules influenced the morphology of the inorganic structure that lodged it.

2. Materials and Methods

2.1. Extraction of the Genomic DNA

The *Escherichia coli* JM109 culture was left to grow for 8 h under constant shaking (120 rpm) in Luria–Bertani medium (LB: 5 g/L yeast extract, 10 g/L tryptone, 5 g/L NaCl). From this culture, 5 mL were taken, and cells were collected by centrifugation at 3000× *g* for 10 min. Then, cells were resuspended in 500 µL of lysis buffer (20 mM Tris-Cl, pH 8.0, 2 mM sodium EDTA, 1.2% Triton X-100, and lysozyme to 20 mg/mL). Starting with the cell lysate, the protocol of the One-4-all genomic DNA kit (Bio Basic Inc., Toronto, ON, Canada) was followed. Briefly, 180 µL of the ACL buffer and 20 µL of proteinase K were added and vortexed. The mixture was incubated in a water bath at 56 °C for 60 min. Then, 200 µL of RNAse- and DNAse-free ethanol at 96% was added. The mixture was transferred to the EZ-10 column and centrifuged at 9000× *g* for 1 min. The supernatant was discarded, 500 µL of solution CW1 was added, and the mixture centrifuged again at 9000× *g* for 1 min; then, 500 µL of solution CW2 was added and centrifuged at 9000× *g* for 1 min. Finally, the DNA was resuspended in 50 µL of nuclease-free water, incubated for 5 min in a water bath at 60 °C, and left to cool at room temperature. The obtained DNA was stored at −20 °C until used in the synthesis of biomorphs. The integrity of DNA was verified in a 0.8% agarose gel. The quantification and purity of the DNA were determined through spectrophotometry at 260 and 280 nm (Nanodrop 2000, Thermo Fisher Scientific, Inc., Waltham, MA, USA), as indicated below.

2.2. Electrophoretic Analysis

The obtained DNA was visualized through electrophoresis of a denaturalized 0.8% agarose gel in TAE 1× buffer (Tris, acetic acid, and EDTA). The gel was stained with 0.1% ethidium bromide, and bands were observed in an UV transilluminator (Gel Doc XR System. Bio-Rad, Hercules, CA, USA).

2.3. Spectrophotometric Analysis

The quality and quantity of the obtained DNA were determined in a spectrophotometer (Nanodrop, 2000, Thermo Fisher Scientific, Inc.). The absorbances at 260 and 280 nm were measured in each sample to estimate their quality ratio at 260 nm/280 nm.

2.4. Biomorphs Formation

The formation of barium silica-carbonate biomorphs was performed by means of the gas diffusion method [22,31]. Experiments were performed on glass 5 mm in length, 5 mm in width, and 1 mm in thickness. The glass plate was placed inside a crystallization cell with a final volume of 200 µL. The solution for the synthesis of biomorphs was prepared with a mixture of 1000 ppm sodium metasilicate, 20 mM barium chloride, and 1.0, 0.5, 0.25, 0.14, and 0.01 ng of genomic DNA. Finally, the pH of the mixture was adjusted to 11.0 with sodium hydroxide. All reagents were from Sigma-Aldrich (St. Louis, MO, USA). Experiments were performed in two different conditions, at a constant 5% CO_2 flux in a CO_2 incubator (NuAire, Plymouth, MN, US), and with CO_2 in standard conditions (STP). In this way, 12 different conditions for biomorph synthesis were obtained. Biomorph formation was allowed for 24 h.

2.5. Characterization of Biomorphs

Biomorphs were observed through scanning electron microscopy (SEM) and analyzed through Raman and Fourier-transform infrared (FTIR) spectroscopy.

2.5.1. Scanning Electron Microscopy (SEM)

Biomorphs were observed by means of SEM microphotographs, using a TESCAN microscope (Brno, Czech Republic) model VEGA3 SB, with a secondary electron detector (SE) from 10 to 20 kV in high vacuum conditions (work distance of 10 mm).

2.5.2. Raman Microspectroscopy

Raman spectra were collected using a WITec alpha300 RA spectrometer (WITec GmbH, Ulm, Germany) under ambient conditions with 532 nm laser light excitation, from a Nd:YVO_4 incident laser beam, with a power of 6.37 mW and detection of 672 lines/mm grating. The incident laser beam was focused by 20×, 50×, and 100× objectives (Zeiss, Oberkochen, Germany) with 0.4, 0.75, and 0.9 NA, respectively.

Punctual Raman spectra were obtained with 0.5 s of integration time and 0.03 s for image mapping. The data processing and analysis were performed with the WITec Project Version 5.1 software.

2.5.3. Fourier-Transform Infrared Spectroscopy (FTIR)

Fourier transform infrared spectroscopy (FTIR) analyses were conducted using a Nicolet iS50R Thermo Scientific spectrometer, equipped with an attenuated total reflectance (ATR) diamond crystal accessory (Smart-iTX). Spectra acquisitions were collected with 32 scans and 4 cm^{-1} spectral resolution, in the range of 525 to 4000 cm^{-1}. The data processing and analysis were performed with the OMNIC version 9 software.

3. Results and Discussion

According to the prevailing conditions on primigenial Earth, we now know that those conditions favored the start of life in our planet. The nucleic acids were the biomolecules from which the cell originated, thus leading to life on Earth. Although the latter is fascinating, we still need to determine the minimal concentration required for the DNA of the pioneer cell to control the morphology that this prebiotic cell had to adopt. Therefore, aimed at knowing whether DNA, even at low concentrations, could affect the morphology of biomorph structures in which DNA became isolated from the environment, we synthetized barium biomorphs at five different DNA concentrations (1.0, 0.5, 0.25, 0.14, and 0.01 ng/µL) and at two CO_2 concentrations, one emulating the Precambrian era (5%) and the other emulating the present one (STP). As observed in Figure 1A, in the control samples at 5% CO_2, biomorphs presented sphere-type and stem structure morphologies. The formation of spheres at high concentrations of DNA and CO_2 was likely caused by the highly charged molecules of the DNA (working as nucleation centers) and the local reduction in pH values at the high concentration of CO_2. On the other hand, in standard CO_2 conditions (STP), morphologies of leaves, stems, and flowers were found (Figure 1B). These morphologies are the most typically reported for biomorphs of barium silica carbonates in these conditions without DNA [21–23,32,33]. Biomorphs obtained at a DNA concentration of 0.01 ng/µL, in both CO_2 conditions, presented almost the same morphology as that of the respective control (Figure 1A–D). Interestingly, at a DNA concentration of 0.14 ng/µL under a 5% CO_2 flux, the morphology was a bit similar to that observed in the control biomorphs (Figure 1A,E). On the other hand, biomorphs with this DNA concentration under STP conditions started to form complex arrangements of helices (Figure 1F). The biomorphs produced at a DNA concentration of 0.25 ng/µL with 5% CO_2 revealed stem morphologies (Figure 1G).

Biomorphs produced under STP conditions were arranged in leaves, short helices, and ribbons (Figure 1H). The biomorphs synthetized at 5% CO_2 at a DNA concentration of 0.5 ng/µL presented structural arrangements such as stems with some spheres or leaves with spheres (Figure 1I). On the other hand, these biomorphs at the same DNA concentration but under STP conditions showed morphologies of complex arrangements such as helices and small spheres at the end (Figure 1J). These morphologies with 0.5 ng/µL DNA were different from the control biomorphs (Figure 1A,B). For the biomorphs synthetized with 1.0 ng/µL at 5% CO_2 (Figure 1K), a spheric morphology was generally observed. This morphology has been adopted and reported by biomorphs in some specific conditions, such as high CO_2 concentrations and in the presence of proteins [29,30]. This result reveals that, at these CO_2 and DNA concentrations, DNA did indeed influence the morphology of the crystals of the obtained biomorphs. Biomorphs obtained at the same DNA concentration but in STP conditions had worm-type structure morphologies, forming arrangements with several of these structures (Figure 1L). The morphology adopted by biomorphs synthetized at DNA concentration of 1.0 ng/µL (Figure 1K,L) was completely different from that obtained at the same CO_2 concentrations (Figure 1A,B). These results reveal that a DNA concentration of 0.01 ng/µL was no longer sufficient to influence the morphology.

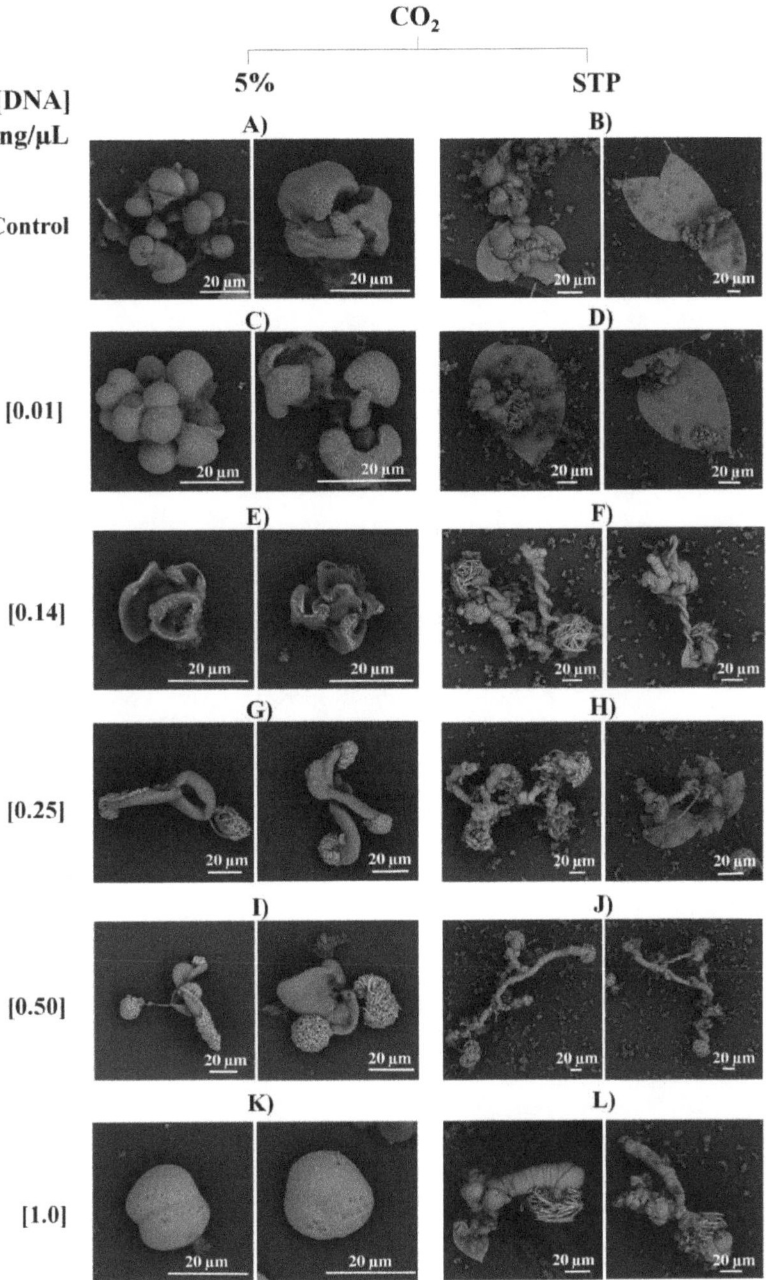

Figure 1. SEM microphotographs of biomorphs synthetized at different DNA concentrations, in the presence of a 5% CO_2 current (**A**,**C**,**E**,**G**,**I**,**K**) or in standard STP conditions (**B**,**D**,**F**,**H**,**J**,**L**).

The chemical composition and the crystalline structure of the obtained biomorphs were determined through Raman and IR spectroscopy. Raman analysis of the control biomorphs at 5% CO_2 identified bands at 93, 138, 223, 690, and 1058 cm^{-1} (Table 1), whereas, in the control biomorphs but under STP conditions, bands at 140, 698, 1059, and

2893 cm^{-1} were identified (Table 1). In both samples, the identified peaks corresponded to the $BaCO_3$ polymorph, aragonite type, named witherite [34]. The microstructure of the barium silica-carbonate crystals was also analyzed through IR spectroscopy, identifying peaks at 629, 787, 855, 937, 1059, 1417, and 1732 cm^{-1} (Table 1). Results confirm that this was indeed witherite.

Table 1. Identification through Raman and IR spectroscopy of the polymorphs of the barium silica-carbonate biomorphs.

Sample/DNA Concentration [ng/μL]	Synthesis Condition	Raman (cm^{-1})	IR (cm^{-1})	Composition
Control (-)	5% CO_2	93, 138, 223, 690, 1058	629, 787, 855, 937, 1059, 1417, 1732	Witherite/abiotic
	STP	140, 698, 1059, 2893	692, 796, 855, 958, 1099, 1417	Witherite/abiotic
[0.01]	5% CO_2	106, 147, 151, 700, 1067, 2915	692, 789, 855, 1059, 1417, 1999	Witherite/abiotic
	STP	93, 138, 691, 1058	692, 787, 796, 855, 946, 1050, 1072, 1416, 1577	Witherite/abiotic
[0.14]	5% CO_2	112, 499, 1456, 2918	649, 787, 937, 1417, 2000	Witherite/abiotic
	STP	141, 155, 693, 1061	592, 692, 767, 899, 1069, 1417, 1577, 2851	Witherite/abiotic
[0.25]	5% CO_2	93, 138, 155, 222, 690, 1058, 2908	582, 692, 789, 855, 1059, 1415, 1732, 1969	Witherite/abiotic
	STP	141, 152, 223, 494693, 1061, 2909, 2967	603, 796, 882, 997, 1075, 1415, 1590, 1749, 1974	Witherite/abiotic
[0.50]	5% CO_2	99, 485, 1280, 1460, 2411, 2910, 2970	693, 787, 856, 891, 937, 1057, 1423, 1732, 1970, 2400, 2850, 2940	Witherite/biotic
	STP	96, 156, 692, 1060, 1360, 1415, 2917	583, 692, 788, 796, 854, 947, 1071, 1416, 1560, 1770, 1979, 2160, 2480, 2820	Witherite/biotic
[1.00]	5% CO_2	100, 498, 893, 1451, 1600, 2911, 2962	629, 761, 787, 891, 1033, 1417, 1732, 1969, 2860, 2929	Witherite/biotic
	STP	96, 150, 224, 691, 1059, 1362, 1421, 2942	608, 692, 796, 854, 968, 999, 1413, 1780, 2470, 2880, 2920	Witherite/biotic

In the biomorphs synthetized at 5% CO_2 in the presence of DNA at a concentration of 1.0 ng/μL, the Raman spectrum revealed bands at 100, 498, 893, 1451, 1600, 2911, and 2962 cm^{-1}, and the IR spectrum revealed peaks at 629, 761, 787, 891, 1033, 1417, 1732, 1969, 2860, and 2929 cm^{-1} (Table 1). Figure 2 shows the most representative images of the biomorphs obtained using scanning electron microscopy (Figure 2A) and optical microscopy (Figure 2B), while Figure 2C corresponds to the Raman spectrum (the inset shows the mapping of one of the biomorphs; Figure 2D).

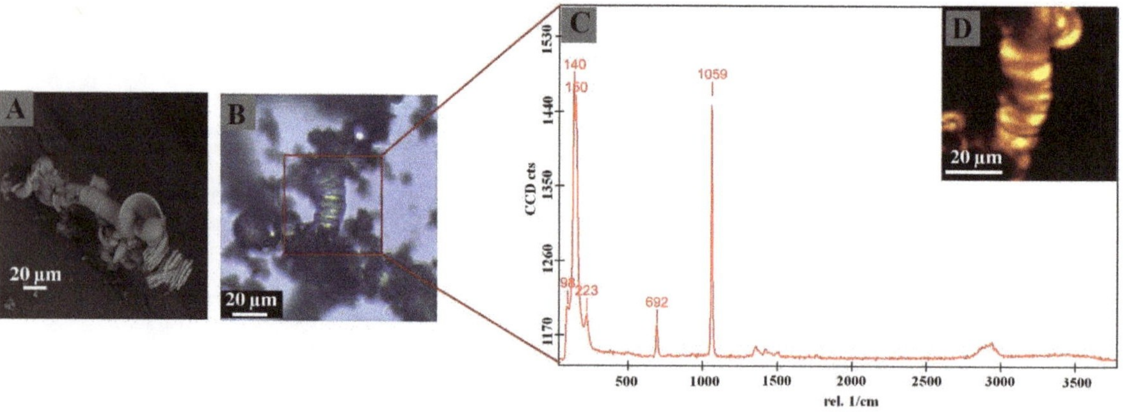

Figure 2. Representative image for the identification of the crystalline phase of $BaCO_3$ biomorphs synthesized at STP conditions at a concentration of DNA of 1.0 ng/μL through Raman spectroscopy: (**A**) SEM microphotograph; (**B**) optical image; (**C**) Raman spectrum; (**D**) mapping of biomorph.

Both Raman and IR peaks were identified as corresponding to the kerogen signal, which has been proposed as a marker of biogenicity [35–38], and our research team identified the biogenic kerogen signal for the first time in biomorphs synthetized in the presence of DNA pertaining to the five kingdoms in nature [29]. In the Raman spectrum, the kerogen signal is identified in two bands around 1300 (Band "D") and 1600 (Band "G") cm^{-1}, and, in the spectrum, there may or may not appear two poorly intense bands between 2600 and 2900 cm^{-1} [37,38]. In biomorphs, bands D and G appear between 1300 and 1700 cm^{-1}; generally, one or two poorly intense bands can also be identified between 2600 and 2900 cm^{-1} [29]. In the IR spectrum, the peaks corresponding to the kerogen signal appear approximately at 700, 900, 1630, 1710, 2890, and 2930 cm^{-1}; there could be variations in the values of the peaks, due to the type of kerogen and its maturation degree [39]. Having identified bands characteristic of kerogen, the samples were identified as biotic type to distinguish them from the biomorphs where the kerogen signal was not identified (Table 1). In biomorphs obtained at a DNA concentration of 1.0 ng/µL, but in STP conditions, the characteristic peaks of witherite with kerogen were identified (Table 1), just like in those crystals of barium silica-carbonate produced at the same DNA concentration but under 5% CO_2. Biomorphs synthetized with DNA at a concentration of 0.5 ng/µL in both CO_2 conditions showed bands corresponding to witherite and kerogen; thus, they were also biotic-type biomorphs (Table 1). On the other hand, in biomorphs obtained at DNA concentrations of 0.25, 0.14, and 0.01 ng/µL under both CO_2 conditions, in both Raman and IR spectra, only witherite bands were identified but not those of kerogen (Table 1). These results are relevant, on the one hand, because, in order to find evidence of life, in the Precambrian cherts, for example, we need to analyze a sample containing a minimal concentration of 0.5 ng. This could be a plausible explanation for why most evidence of unicellular life in the primigenial era was lost. In biomorphs, for example, these are generally found on nanometric scales, emulating somehow the unicellular life of the Precambrian. On the other hand, it shows that, even at a given DNA concentration of 0.25 ng, DNA still directs the morphology of the biomorph. That concentration, however, is nonetheless insufficient to identify the trace of a biomarker, such as kerogen. It has been proposed that, in different organisms, the DNA concentration varies from 0.002 pg in prokaryotes to 700 pg in eukaryotes [40–42]. This DNA concentration that can be identified in the cells of prokaryotes or eukaryotes differs from the DNA concentration used in the biomorphs; for example, in this case, the kerogen marker was identified in the different obtained biomorphs, which is why biomorphs have been suggested as the antecessors of the Precambrian cherts [29]. Hence, the kerogen must be identified in the biomorphs as a biomarker rather than a determinant of the DNA concentration. The DNA concentration is related to phenotypical characteristics such as the nuclear and cellular volume, the time of duplication, and the rate of embryonic development [40]. Hence, in prokaryote cells, when there is a 100% increase of DNA content, there is a 1060% increase in the cellular volume, whereas, in unicellular eukaryotes, if DNA increases by 100%, the increase in cellular volume is 93% [40]. This variation in both cellular types reveals that the DNA content generates structural differences, which we also found in biomorphs, where the DNA concentration changed the morphology together with the environmental conditions (Figure 1). This is possibly the reason why, in the Precambrian era, the first forms of life adopted morphologies in spheric shapes, with tips and stems, similarly to the morphologies adopted in conditions that emulated the Precambrian era (Figures 1 and 3).

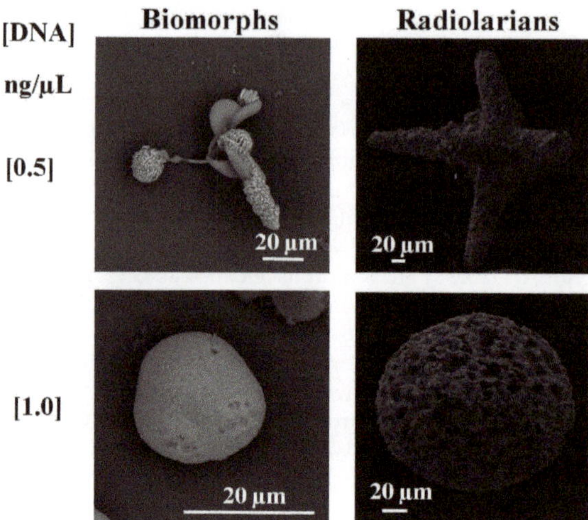

Figure 3. Representative microphotographs of biomorphs synthetized at 1.0 and 0.5 ng/µL of DNA with 5% CO_2 in comparison with a radiolarian of sponge spicules from the Niujiaohe Formation in Chongyi County, Jiangxi Province (with permission from Zhang and Feng, 2019 [43]).

The DNA content in higher organisms has also been shown to vary not only among organisms, but also differs among different cell types of the same organism. Thus, in a study that determined the DNA concentration of 19 different human cell types, it was shown that the amount of DNA was related to the increase in cellular volume [44]. The DNA content in humans compared with other vertebrates varies, e.g., 2–448 pg in human cells, 1–266 pg in fishes, 2–260 pg in amphibians, 2–11 pg in reptiles, 3–17 pg in mammals, and 2–4 pg in birds (Figure 4) [44]. It is important to remark that the vertebrate animals include humans; the plots in Figure 4A and, particularly, Figure 4B show the content of DNA in the cell volume in different types of genomes.

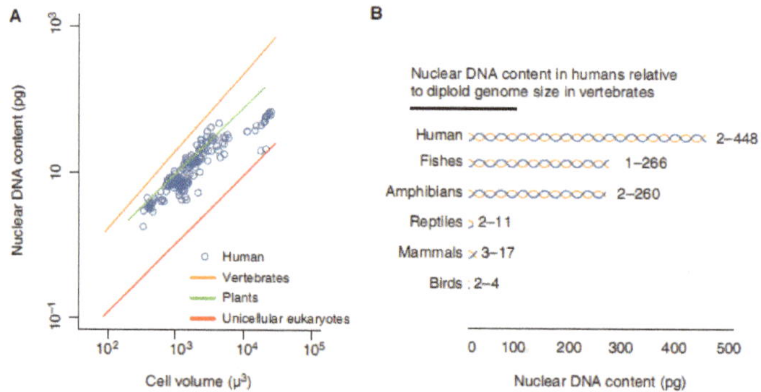

Figure 4. Relationship between nuclear DNA content and cell volume in humans. (**A**) Relationship between nuclear DNA content and cell size in diploid and polyploid human cells (blue) in comparison to previously reported relationships for diploid cells of vertebrates, unicellular eukaryotes, and angiosperms. Lines were fitted to previously reported relationships using ordinary least-squares regression. The range of the unicellular eukaryote cell and genome sizes is truncated to clearly show data for other groups. (**B**) Range of nuclear DNA content in individual human cells in comparison to ranges of diploid genome sizes within vertebrate groups. DNA content was rounded to the nearest whole number; plot and data were taken from [44] with copyright permission.

The relationship between the nuclear DNA concentration and the size of the cell has also been observed in plants, in which we identified that the increase in DNA content is associated with cellular differentiation [45]. In plants, the amount of DNA differs among the different organelles and is associated with the cellular size, while the nuclear DNA content varies among species of the same genus [46–48]. Additionally, several authors have associated environmental factors (e.g., temperature, elevation, latitude, and precipitation) with the DNA concentration in the cells of plants [49–52].

This observation that the DNA concentration in plants depends also on environmental factors can be correlated with our results, because we found that the morphology adopted by the biomorphs varied depending on the CO_2 concentration, favoring the morphology of leaves and flowers in STP conditions (Figures 1 and 3), indicating that this morphology is favored in the current environment. On the other hand, the fact that the amount of DNA increases due to environmental conditions may explain why, in the current conditions of our atmosphere, the DNA could have increased in the first organisms, leading to an increment in the cellular size and, thus, giving rise to more complex organisms. The fact that DNA is the main biomolecule to which the cellular size is attributed is not a trivial issue because the understanding of what contributes to cell size has been evaluated for more than one century [38].

4. Conclusions

Synthesis of nucleic acids in the Precambrian era of the Earth marked the start of life, with DNA being the molecule where genetic information has been preserved ever since. In this work, we show for the first time how the DNA at relatively low concentrations is able to influence the morphology of the structure in which it is found, e.g., in biomorphs. Our results indicate that, even at a low DNA concentration, as was the case since the synthesis of the first DNA biomolecules available in the primigenial era, these biomolecules influenced the morphology of the inorganic structure in which they were found. This allowed the DNA to become polymerized and increase its concentration, leading to an increase in the size of the first cells, giving origin to cells with different morphologies, and leading to the formation of more complex organisms. On the other side, results allow inferring that, since the Precambrian era, once the DNA was synthetized, it has been responsible for conserving and directing the morphology of all organisms until our current time.

Author Contributions: Conceptualization, M.C.-C. and A.M.; methodology, C.D.P.-A., M.C.-C. and S.R.I.; software and validation, M.C.-C. and S.R.I.; formal analysis, M.C.-C. and A.M.; investigation, C.D.P.-A., M.C.-C., S.R.I. and A.M.; writing—original draft preparation, review, and editing, M.C.-C. and A.M. All authors have read and agreed to the published version of the manuscript.

Funding: This work was carried out with the financial support granted to M. Cuéllar-Cruz by Project No. CF2019-39216 from the *Consejo Nacional de Ciencia y Tecnología* (CONACYT) and *Proyecto-Institucional-UGTO*-017/2022 from the *Universidad de Guanajuato*, Mexico. Additionally, this work was partially supported by the project CONACYT No. A1-S-7509.

Acknowledgments: This work was carried out with the financial support granted to M. Cuéllar-Cruz by Project No. CF2019-39216, as well as Project No. A1-S-7509 granted to Abel Moreno from the *Consejo Nacional de Ciencia y Tecnología* (CONACYT) and *Proyecto-Institucional-UGTO*-017/2022 from the *Universidad de Guanajuato*, Mexico. The authors acknowledge Ingrid Mascher for the English revision of this manuscript, as well as Antonia Sánchez Marín for the final revision and English style correction of the revised manuscript. Cesia D. Pérez-Aguilar acknowledges the *Proyecto-Institucional-UGTO*-017/2022 grant from the *Universidad de Guanajuato*, Mexico. The authors thank the Laboratorio Universitario de Caracterización Espectroscópica, LUCE-ICAT-UNAM, and José Guadalupe Bañuelos for their support with the Raman measurements.

Conflicts of Interest: The authors declare no conflict of interest.

References

1. Oro, J. Chemical evolution and the origin of life. *Adv. Space Res.* **1983**, *9*, 77–94. [CrossRef]
2. Powner, M.; Gerland, B.; Sutherland, J. Synthesis of activated pyrimidine ribonucleotides in prebiotically plausible conditions. *Nature* **2009**, *459*, 239–242. [CrossRef] [PubMed]
3. Petrov, A.S.; Gulen, B.; Norris, A.M.; Kovacs, N.A.; Bernier, C.R.; Lanier, K.A.; Fox, G.E.; Harvey, S.C.; Wartell, R.M.; Hud, N.; et al. History of the ribosome and the origin of translation. *Proc. Natl. Acad. Sci. USA* **2015**, *112*, 15396–15401. [CrossRef]
4. Ban, N.; Nissen, P.; Hansen, J.; Moore, P.B.; Steitz, T.A. The complete atomic structure of the large ribosomal subunit at 2.4 Å resolution. *Science* **2000**, *289*, 905–920. [CrossRef]
5. Fedor, M.J.; Williamson, J.R. The catalytic diversity of RNAs. *Nat. Rev. Mol. Cell Biol.* **2005**, *6*, 399–412. [CrossRef]
6. Brunk, C.F.; Marshall, C.R. Whole Organism, Systems Biology, and Top-Down Criteria for Evaluating Scenarios for the Origin of Life. *Life* **2021**, *11*, 690. [CrossRef]
7. Yadav, M.; Kumar, R.; Krishnamurthy, R. Chemistry of Abiotic Nucleotide Synthesis. *Chem. Rev.* **2020**, *120*, 4766–4805. [CrossRef]
8. Orgel, L.E.; Lohrmann, R. Prebiotic Chemistry and Nucleic Acid Replication. *Acc. Chem. Res.* **1974**, *7*, 368–377. [CrossRef]
9. Lazcano, A.; Miller, S.L. The Origin and Early Evolution of Life: Prebiotic Chemistry, the Pre-RNA World, and Time. *Cell* **1996**, *85*, 793–798. [CrossRef]
10. Oro, J.; Stephen-Sherwood, E. The Prebiotic Synthesis of Oligonucleotides. In *Cosmochemical Evolution and the Origins of Life: Proceedings of the Fourth International Conference on the Origin of Life and the First Meeting of the International Society for the Study of the Origin of Life, Barcelona, June 25—28, 1973, Volume I: Invited Papers and Volume II: Contributed Papers*; Oró, J., Miller, S.L., Ponnamperuma, C., Young, R.S., Eds.; Springer Netherlands: Dordrecht, The Netherlands, 1974; pp. 159–172.
11. Powner, M.W.; Zheng, S.L.; Szostak, J.W. Multicomponent Assembly of Proposed DNA Precursors in Water. *J. Am. Chem. Soc.* **2012**, *134*, 13889–13895. [CrossRef]
12. Follmann, H. Deoxyribonucleotides: The Unusual Chemistry and Biochemistry of DNA Precursors. *Chem. Soc. Rev.* **2004**, *33*, 225–233. [CrossRef] [PubMed]
13. Wachtershauser, G. The Place of RNA in the Origin and Early Evolution of the Genetic Machinery. *Life* **2014**, *4*, 1050–1091. [CrossRef] [PubMed]
14. Dortch, Q.; Roberts, T.; Clayton, J.; Ahmed, S. RNA/DNA ratios and DNA concentrations as indicators of growth rate and biomass in planktonic marine organisms. *Mar. Ecol. Prog. Ser.* **1983**, *13*, 61–71. [CrossRef]
15. Caldwell, P.C.; Hinshelwood, C. The nucleic acid content of *Bact. lactis aerogenes*. *J. Chem. Soc.* **1950**, 1415–1418. [CrossRef]
16. Dennis, P.P.; Bremer, H. Macromolecular composition during steady-state growth of Escherichia coli B-r. *J Bacteriol.* **1974**, *119*, 270–281. [CrossRef] [PubMed]
17. Chícharo, M.A.; Chícharo, L. RNA:DNA ratio and other nucleic acid derived indices in marine ecology. *Int. J. Mol. Sci.* **2008**, *9*, 1453–1471. [CrossRef]
18. Leick, V. Ratios between contents of DNA, RNA and protein in different micro-organisms as a function of maximal growth rate. *Nature* **1968**, *217*, 1153–1155. [CrossRef]
19. Mirsky, A.; Ris, H. Variable and Constant Components of Chromosomes. *Nature* **1949**, *163*, 666–667. [CrossRef]
20. Wächtershäuser, G. From volcanic origins of chemoautotrophic life to Bacteria, Archaea and Eukarya. *Philos. Trans. R. Soc. Lond. B Biol. Sci.* **2006**, *361*, 1787–1808. [CrossRef]
21. Cuéllar-Cruz, M.; Moreno, A. Synthesis of crystalline silica-carbonate biomorphs of Ba (II) under the presence of RNA and positively and negatively charged ITO electrodes: Obtainment of graphite via bioreduction of CO_2 and its implications to the chemical origin of life on primitive Earth. *ACS Omega* **2020**, *5*, 5460–5469.
22. Cuéllar-Cruz, M.; Islas, S.R.; González, G.; Moreno, A. Influence of nucleic acids on the synthesis of crystalline Ca (II), Ba (II), and Sr (II) silica-carbonate biomorphs: Implications for the chemical origin of life on primitive Earth. *Cryst. Growth Des.* **2019**, *19*, 4667–4682. [CrossRef]
23. Cuéllar-Cruz, M.; Moreno, A. The role of calcium and strontium as the most dominant elements during combinations of different alkaline Earth metals in the synthesis of crystalline silica-carbonate biomorphs. *Crystals.* **2019**, *9*, 381. [CrossRef]
24. Cuéllar-Cruz, M.; Scheneider, D.K.; Stojanoff, V.; Islas, S.R.; Sánchez-Puig, N.; Arreguín-Espinosa, R.; Delgado, J.M.; Moreno, A. Formation of crystalline silica-carbonate biomorphs of alkaline Earth metals (Ca, Ba, Sr) from ambient to low temperatures: Chemical implications during the primitive Earth's life. *Cryst. Growth Des.* **2020**, *20*, 1186–1195. [CrossRef]
25. Cuéllar-Cruz, M. Influence of abiotic factors in the chemical origin of life: Biomorphs as a study model. *ACS Omega* **2021**, *6*, 8754–8763. [CrossRef]
26. Zhang, G.; Morales, J.; Garcia-Ruiz, J.M. Growth behaviour of silica/carbonate nanocrystalline composites of calcite and aragonite. *J. Mater. Chem. B* **2017**, *5*, 1658–1663. [CrossRef] [PubMed]
27. García-Ruiz, J.M.; Hyde, S.T.; Carnerup, A.M.; Christy, A.G.; Kranendonk, V.M.J.; Welham, N.J. Self-Assembled Silica-Carbons structures and detection of ancient microfossils. *Science* **2003**, *302*, 1194–1197. [CrossRef] [PubMed]
28. Opel, J.; Wimmer, F.P.; Kellermeier, M.; Colfen, H. Functionalisation of silica-carbonate biomorphs. *Nanoscale Horiz.* **2016**, *1*, 144–149. [CrossRef]
29. Cuéllar-Cruz, M.; Islas, S.R.; Ramírez-Ramírez, N.; Pedraza-Reyes, M.; Moreno, A. Protection of the DNA from selected species of five kingdoms in Nature by Ba(II), Sr(II), and Ca(II) silica-carbonates: Implications about biogenicity and evolving from the prebiotic chemistry to biological chemistry. *ACS Omega* **2022**. *submitted at August 2022*.

30. Sánchez-Puig, N.; Cuéllar-Cruz, M.; Islas, S.R.; Tapia-Vieyra, J.V.; Arreguín-Espinosa, R.A.; Moreno, A. The influence of silicateins on the shape and crystalline habit of silica carbonate biomorphs of alkaline Earth metals (Ca, Ba, Sr). *Cyrstals* **2021**, *11*, 438. [CrossRef]
31. Noorduin, W.L.; Grinthal, A.; Mahadevan, L.; Aiznberg, J. Rationally Designed Complex, Hierarchical Microarchitectures. *Science* **2013**, *340*, 832–837. [CrossRef]
32. García-Ruiz, J.M.; Melero-García, E.; Hyde, S.T. Morphogenesis of self-assembled nanocrystalline materials of barium carbonate and silica. *Science* **2009**, *362*, 362–365. [CrossRef]
33. Montalti, M.; Zhang, G.; Genovese, D.; Morales, J.; Kellermeier, M.; Garcia-Ruiz, J.M. Local pH oscillations witness autocatalytic self organization of biomorphic nanostructures. *Nat. Commun.* **2017**, *8*, 14427. [CrossRef] [PubMed]
34. Lin, C.C.; Liu, L.G. High-pressure Raman spectroscopic study of post-aragonite phase transition in witherite ($BaCO_3$). *Eur. J. Miner.* **1997**, *9*, 785–792. [CrossRef]
35. Strauss, H.; Moore, T.B. Abundances and isotopic compositions of carbon and sulfur species in whole rock and kerogen samples. In *The Proterozoic Biosphere, A Multidisciplinary Study*; Schopf, J.W., Klein, C., Eds.; Cambridge University Press: New York, NY, USA, 1992; pp. 709–798.
36. Schopf, J.W.; Kudryavtsev, A.B. Biogenicity of Earth´s earliest fossils: A resolution of the controversy. *Gondwana Res.* **2012**, *22*, 761–771. [CrossRef]
37. Schopf, J.W.; Kudryavtsev, A.B.; Agresti, D.G.; Wdowiak, T.J.; Czaja, A.D. Laser-Raman imagery of Earth's earliest fossils. *Nature* **2002**, *416*, 73–76. [CrossRef]
38. Marshall, W.F.; Young, K.D.; Swaffer, M.; Wood, E.; Nurse, P.; Kimura, A.; Frankel, J.; Wallingford, J.; Walbot, V.; Qu, X.; et al. What determines cell size? *BMC Biol.* **2012**, *10*, 101. [CrossRef] [PubMed]
39. Ganz, H.H.; Kalkreuth, W. IR classification of kerogen type, thermal maturation, hydrocarbon potential and lithological characteristics. *J. Southeast Asian Earth Sci.* **1991**, *5*, 19–28. [CrossRef]
40. Shuter, B.J.; Thomas, J.E.; Taylor, W.D.; Zimmerman, A.M. Phenotypic correlates of genomic DNA content in unicellular eukaryotes and other cells. *Am. Nat.* **1983**, *122*, 26–44. [CrossRef]
41. Gunge, N.; Nakatomi, Y. Genetic mechanisms of rare matings of the yeast Saccharomyces cerevisiae heterozygous for mating type. *Genetics* **1972**, *70*, 41–58. [CrossRef]
42. Pedersen, R.A. DNA content, ribosomal gene multiplicity and cell size in fish. *J. Exp. Zool.* **1971**, *177*, 65–78. [CrossRef]
43. Zhang, K.; Feng, Q.L. Early Cambrian radiolarians and sponge spicules from the Niujiaohe Formation in South China. *Palaeoworld* **2019**, *28*, 234–242. [CrossRef]
44. Gillooly, J.F.; Hein, A.; Damiani, R. Nuclear DNA Content Varies with Cell Size across Human Cell Types. *Cold Spring Harb. Perspect. Biol.* **2015**, *7*, a019091. [CrossRef] [PubMed]
45. Zhang, C.; Gong, F.C.; Lambert, G.M.; Galbraith, D. Cell type-specific characterization of nuclear DNA contents within complex tissues and organs. *Plant Methods* **2005**, *1*, 7. [CrossRef] [PubMed]
46. Rauwolf, U.; Golczyk, H.; Greiner, S.; Herrmann, R.G. Variable amounts of DNA related to the size of chloroplasts III. Biochemical determinations of DNA amounts per organelle. *Mol. Genet. Genom.* **2010**, *283*, 35–47. [CrossRef] [PubMed]
47. Jovtchev, G.; Schubert, V.; Meister, A.; Barow, M.; Schubert, I. Nuclear DNA content and nuclear and cell volume are positively correlated in angiosperms. *Cytogenet. Genome Res.* **2006**, *114*, 77–82. [CrossRef] [PubMed]
48. Hendrix, B.; Stewart, J.M. Estimation of the nuclear DNA content of gossypium species. *Ann. Bot.* **2005**, *95*, 789–797. [CrossRef]
49. Bennett, M.D.; Smith, J.D.; Lewis-Smith, R.I. DNA amounts of angiosperms from the Antartic and South Georgia. *Environ. Exp. Bot.* **1982**, *22*, 307–318. [CrossRef]
50. Caceres, M.E.; Pace, C.D.; Mugnozza, G.T.S.; Kotsonis, P.; Ceccarelli, M.; Cionini, P.G. Genomes size variations within Dasypyrum villosum: Correlations with chromosomal traits, environmental factors and plant phenotypic characteristics and behavior in reproduction. *Theoret. Appl. Genet.* **1998**, *96*, 559–567. [CrossRef]
51. Rayburn, A.L. Genome size variation in Southwestern United States Indian maize adapted to various altitudes. *Evol. Trends Plants* **1990**, *4*, 53–57.
52. Wakamiya, I.; Newton, R.J.; Johnston, S.J.; Price, J.H. Genome size and environmental factors in the genus Pinus. *Am. J. Bot.* **1993**, *80*, 1235–1241. [CrossRef]

Article

Through Diffusion Measurements of Molecules to a Numerical Model for Protein Crystallization in Viscous Polyethylene Glycol Solution

Hiroaki Tanaka [1], Rei Utata [2], Keiko Tsuganezawa [2,3], Sachiko Takahashi [1] and Akiko Tanaka [1,2,3,*]

1 Confocal Science Inc., Musashino Bld. 5-14-15 Fukasawa, Tokyo 158-0081, Japan; tanakah@confsci.co.jp (H.T.); takahashis@confsci.co.jp (S.T.)
2 RIKEN Systems and Structural Biology Center, 1-7-22 Suehiro-cho, Tsurumi, Yokohama 230-0045, Japan; keiko.tsuganezawa@riken.jp (K.T.)
3 RIKEN Center for Biosystems Dynamics Research, 1-7-22 Suehiro-cho, Tsurumi, Yokohama 230-0045, Japan
* Correspondence: aktanaka@a.riken.jp

Abstract: Protein crystallography has become a popular method for biochemists, but obtaining high-quality protein crystals for precise structural analysis and larger ones for neutron analysis requires further technical progress. Many studies have noted the importance of solvent viscosity for the probability of crystal nucleation and for mass transportation; therefore, in this paper, we have reported on experimental results and simulation studies regarding the use of viscous polyethylene glycol (PEG) solvents for protein crystals. We investigated the diffusion rates of proteins, peptides, and small molecules in viscous PEG solvents using fluorescence correlation spectroscopy. In high-molecular-weight PEG solutions (molecular weights: 10,000 and 20,000), solute diffusion showed deviations, with a faster diffusion than that estimated by the Stokes–Einstein equation. We showed that the extent of the deviation depends on the difference between the molecular sizes of the solute and PEG solvent, and succeeded in creating equations to predict diffusion coefficients in viscous PEG solutions. Using these equations, we have developed a new numerical model of 1D diffusion processes of proteins and precipitants in a counter-diffusion chamber during crystallization processes. Examples of the application of anomalous diffusion in counter-diffusion crystallization are shown by the growth of lysozyme crystals.

Keywords: protein crystallization; nucleation; viscosity; diffusion; PEG; FCS; counter-diffusion; self-searching; crystallization scenario

Citation: Tanaka, H.; Utata, R.; Tsuganezawa, K.; Takahashi, S.; Tanaka, A. Through Diffusion Measurements of Molecules to a Numerical Model for Protein Crystallization in Viscous Polyethylene Glycol Solution. *Crystals* **2022**, *12*, 881. https://doi.org/10.3390/cryst12070881

Academic Editor: Abel Moreno

Received: 30 May 2022
Accepted: 20 June 2022
Published: 21 June 2022

1. Introduction

In protein-crystallization studies, polyethylene glycols (PEGs) are one of the main types of precipitants added to protein solutions to reduce their solubility and produce the supersaturation conditions required for the nucleation and growth of crystals. Many protein structures have been determined by growing protein crystals in solutions containing PEGs of various molecular weights (MW) (200–20,000 g/mol) and various concentrations (up to 40 w/v%) as precipitants [1–3]. McPherson and Gavira reported that PEGs with an MW ranging from 2000 to 8000 are the most useful precipitants, and most protein crystals were grown in solutions containing from 4 to 18% PEG for crystallization [1]. These PEG solutions are viscous, and thus may affect the mass transportation, nucleation and growth processes of the protein crystallization. However, only a few papers have experimentally studied the effect of viscosity on protein crystallization [4,5]. Therefore, we precisely measured diffusion rates in viscous solvents and found the advantages of crystallization conditions with pre-mixing viscous high-MW PEGs in the protein solution and diffusing salt as a second component of the precipitant for counter-diffusion crystallization.

Crystallographic technologies are rapidly changing due to the introduction of automation and high-throughput approaches [6,7]; however, it is necessary to reach a supersaturation state for protein crystallization, so that hundreds of crystallization conditions are often tested to allow for a single protein to acquire diffraction-quality crystals. However, some important proteins are difficult to purify in large quantities. Thus, automated microfluidic systems that produce crystals using a counter-diffusion technique have been developed to save the protein sample and obtain good crystals [8]. Counter-diffusion (also known as liquid–liquid diffusion) is a common crystallization method, in which the protein and precipitants are loaded on opposite sides of a tubing chamber and gradually mixed through their diffusion [3,9,10]. In most cases, the protein is loaded into the capillary, the end of which is sealed with a gel plug to keep the protein from flowing out and allow the precipitant to diffuse in from the reservoir. When the precipitant concentration is sufficiently high and the protein chamber is sufficiently long, precipitation and crystallization of the protein occur along the tubing chamber as a result of self-searching for the optimal crystallization scenario [11]. However, in actual experiments, reagent concentrations and the lengths of the tubing chamber have limitations. Therefore, creating an equation that predicts the diffusive mass transport of proteins and precipitants in the system is crucial to optimize the crystallization conditions of counter-diffusion systems [12,13].

The translational diffusion of molecules in solution is described by the well-known Stokes–Einstein (SE) equation:

$$D = {k_B T}/{6\pi\eta r} \tag{1}$$

where k_B is the Boltzmann constant. The diffusion coefficient (D) depends on the hydrodynamic radius of the molecule approximated as a sphere with radius r, temperature T, and solvent viscosity η. According to the SE equation, the diffusion rates of molecules are inversely proportional to the viscosity of the solvent, although, in some macromolecular crowding solvents, such as high-MW PEGs, the measured diffusion coefficient is larger than a value calculated using the macroviscosity of the solution. The phenomena are known as anomalous diffusion [14,15]. We proposed empirical equations to estimate the D value of a protein in a PEG solution from the protein, PEG MWs and concentration of the PEG [16]. Later, Holyst et al. showed that, when considering the viscosity around the molecules (nanoviscosity), the SE equation is correct, even in viscous PEG solutions [17]. Thus, we planned an experimental study of the anomalous effects of PEG solvents on the diffusion rates of small molecules, peptides, and proteins to predict D values in viscous PEG solvents.

We used fluorescence correlation spectroscopy (FCS) to analyze the diffusion phenomenon. FCS is an effective experimental method to observe the translational diffusion of a molecule under various solvent conditions. The original concept of FCS is to detect and analyze the spontaneous fluctuations in the fluorescence emissions of several molecules in a small detection volume caused by thermodynamic fluctuation [18]. The translational diffusion time (τ_D), in which a molecule remains in the focal volume depending on its hydrodynamic radius, and is proportional to the cubic root of its molecular mass for a spherical particle. Experimentally, τ_D is determined by calculating the autocorrelation function from the fluorescence-intensity fluctuations caused by fluorescent molecules diffusing in and out of the detection volume.

According to Holyst et al. [17], the SE equation can be rewritten as Equation (2). If the diffusion constant of the solute in water is assumed to be D_0 and the viscosity of water η_0, this can be written as:

$$D = {k_B T}/{6\pi\eta_{nano} r} = {D_0 \times \eta_0}/{\eta_{nano}} \tag{2}$$

where D and η_{nano} are the diffusion constant of the solute and the nanoviscosity around the solute molecule in the solvent. The τ_D value of a solute is inversely proportional to D; therefore, Equation (2) shows the following relation:

$$\tau_D = {k_1}/{D} = {k_1 \eta_{nano}}/{D_0 \eta_0} = {k_1 \eta_{nano} \tau_{D0}}/{\eta_0} = k_2 \tau_{D0} \times \eta_{nano} \tag{3}$$

where k_1 and k_2 are specific constants.

We measured the τ_D values of small molecules, peptides, and proteins in various viscous PEG solvents in this paper. In viscous, high-MW-PEG solutions, the diffusion rates of small molecules are selectively susceptible to anomalous diffusion and not inversely proportional to the viscosity of the solution (macroviscosity). Our previous model [16] is greatly improved in this paper with the use of the newly measured results.

The substantial difference between the diffusions of a small molecule and a protein in PEG solution indicated the following benefits to counter-diffusion crystallization in viscous high-MW-PEG solutions: the delays in the diffusion rates of the various precipitants, such as salts and organic solvents, will be smaller than previously expected, and, at the same time, the protein leakage from the crystallization chamber will be suppressed. We have developed a new numerical model of the 1D diffusion processes of the proteins and precipitants in the counter-diffusion chamber, based on the diffusion times of molecules newly measured in the PEG solutions. The new calculation allows us to point out the advantage of using high-MW PEGs in the counter-diffusion crystallization. We have also demonstrated the advantage with an example of lysozyme crystallization.

2. Materials and Methods

2.1. Materials

PEG1000, PEG3350, PEG6000, PEG10,000, and PEG20,000 were purchased from Hampton Research Chemicals (Aliso Viejo, CA, USA). The fluorescent molecules, TAMRA (MW = 528.0) and ALEXA647 (MW = 1155.1), were purchased from Olympus (Tokyo, Japan). The TAMRA-labeled synthetic peptides, SHP-1 binding peptide (10-mer TAMRA: ITpYSLLKGGK-TAMRA, MW = 1572.6) and p53 N-terminal peptide (16-mer TAMRA: SQETFSDLWKLLPEN-K-TAMRA, MW = 2346.6), were purchased from Toray Research Center, Inc. (Tokyo, Japan). Alexa Fluor 647-labelled goat anti-human immunoglobulin G (IgG, MW = 150,000) was purchased from Molecular Probes (Eugene, OR, USA). Egg-white lysozyme was purchased from FUJIFILM (Tokyo, Japan).

2.2. Preparation of the Protein Samples

Human-bromodomain-containing protein 2 (Brd2) (accession: P25440.2, 74aa-194aa, MW = 15,000), and mouse secernin-1 (accession: Q9CZC8.1, 1aa-414aa, MW = 46,300) were synthesized by the Escherichia coli. cell-free protein-synthesis method [19]. The coding cDNA fragments were amplified by polymerase chain reaction, to add a T7 promoter and histidine affinity tag-encoding sequence to the 5′-region, and a T7 terminator sequence to the 3′-region. These subclones were ligated into the pCR2.1-TOPO vector (Invitrogen, Carlsbad, CA, USA) for cell-free protein expression [19–21].

The proteins produced by the *E. coli* cell-free synthesis system with an N-terminal histidine affinity tag with a TEV cleavage site were purified by a HisTrap HP column (Cytiva, Marlborough, MA, USA). The histidine affinity tag was then removed by incubation with TEV protease at 277 K overnight, and the proteins were further purified by a HisTrap HP column and concentrated [22]. Thus, the final protein samples contained additional amino acid sequences derived from the expression vectors, that is, GSSGSSG for the N terminus and SGPSSG for the C terminus. The molecular weights of the final protein samples were 15,388.5 for Brd2 and 47,317.1 for secernin-1.

2.3. Protein Labeling

The amine side chains of the proteins were randomly labeled with Alexa Fluor 647 mono-functional succinimidyl ester (Molecular Probes, Eugene, OR, USA). The Alexa Fluor 647 fluorophore was covalently attached to the protein by the conjugation protocol of the manufacturer. In brief, 13.7 mM protein was dissolved in 0.1 M Na_2CO_3 (pH 8.3) containing the fluorescent dye, and the mixture was stirred for 1 h at room temperature. The free dye molecules were then removed using BioGel P-30 Fine size-exclusion purification resin (Bio-Rad, Hercules, CA, USA). After the labeling step, absorption spectra indicated that 1.26 mole Alexa Fluor was contained in 1 mole Brd2 molecules, and 1.12 mole in 1 mole secernin-1 molecules.

2.4. FCS Measurements

The FCS measurements of the solution samples were performed with an MF20 single-molecule fluorescence detection system (Olympus, Tokyo, Japan) using the on-board 543 nm (for TAMRA) or 633 nm (for Alexa Fluor) helium–neon laser at a laser power of 100 μW for excitation [23,24]. For convenience, the experiments were performed in 384-well glass-bottom plates using a sample volume of 30 μL.

The fluorescent samples and each PEG solution (*w*/*v*) were mixed in FCS buffer containing 50 mM Tris/HCl (pH 8.0) and 0.05% Tween 20. Tween 20 was added to suppress glass–surface interactions. The final concentrations of the fluorescent molecules used for the measurements were adjusted to a concentration of 1 nM. All of the FCS measurements were performed in duplicate. The measurement data were obtained with a data acquisition time of 10 s per measurement, and the measurements were performed five times per sample at 296 K. For machine performance verification and normalization of the obtained results, the τ_D values of the standard fluorescent dyes (1 nM TAMRA or 1 nM Alexa Fluor 647) were determined at each measurement. FCS data analysis was performed with the MF20 software package (Olympus). The error bars shown in the graphs represent the standard deviations of five measurements. The Brd2 and secernin-1 samples were purified after fluorescent labelling, although the samples still contained the free-labeling reagent (29% for the Brd2 sample, and 16% for the secernin-1 sample). Therefore, the τ_D values were analyzed by the two-component analysis software of the instrument (MF20) and decomposed into two τ_D values owing to the protein and contaminating free-labeling reagent (Alexa647). For the analysis, the τ_D value of Alexa647 in each PEG solvent was measured and used.

2.5. Viscosity Measurements

Each PEG solution (*w*/*v*%) containing the FCS buffer was prepared by gentle stirring. The viscosities of the PEG solutions were measured at 296 K by a SV10 sine-wave vibro-viscometer (A&D Company, Ltd., Tokyo, Japan) calibrated with JS10 and JS100 calibration solutions (Nippon Grease Co., Ltd., Yokohama, Japan).

2.6. Crystallization

The gel-tube counter-diffusion methods [13,25] were used, as previously mentioned. A capillary (0.47 mm bore) with a gel tube was filled with 7 μL of a 20% PEG solution (PEG 4000, 10,000, or 20,000) containing 20 mg/mL lysozyme and 50 mM acetate buffer (pH 4.5). The capillaries were vertically placed in a reservoir solution containing 600 mM NaCl, 20% PEG, and 50 mM acetate buffer (pH 4.5) for 30 days at 293 K. Gel tubes with a 1 mm bore were presoaked in each reservoir solution for more than 1 week before sample loading.

2.7. X-ray Diffraction Experiment

The crystals grown in the capillary were picked out and immersed in cryo-protectant solution, including 600 mM NaCl, 40% PEG4000, and 50 mM acetate buffer (pH 4.5) using a cryoloop. They were then flash-frozen with liquid nitrogen.

Data collection was performed using synchrotron radiation at Diamond Light Source beamline i04 equipped with an Eiger2 XE 16M pixel detector. All of the datasets were integrated and scaled using the programs iMosflm [26] and Aimless [27], as implemented in the CCP4 program package [28].

3. Results and Discussion

3.1. Macroviscosity (η_{macro}) of PEG Solutions

The PEG polymers with various MW (1000–20,000) were dissolved in an FCS buffer at 1–20% (*w*/*v*), and their viscosities were measured (Table 1). The high-MW PEGs (PEG10,000 and PEG20,000) strongly increased the viscosities of the solutions and the lower-MW PEGs

(PEG1000 and PEG3350) were not effective, even at 20%. The measured densities of the solutions were 1.01 g/cm^3 (1, 2 and 5%), 1.02 g/cm^3 (10%), 1.03 g/cm^3 (15%), and 1.04 g/cm^3 (20%).

Table 1. Measured macroviscosities of the PEG solutions in the FCS buffer.

MW of PEGs	Viscosity of PEG Solution (mPa·s)					
	Concentrations of PEG Solution * (*w*/*v*%)					
	1	2	5	10	15	20
1000	1.25	1.34	1.58	2.13	2.88	4.00
3350	1.33	1.44	1.99	3.53	5.83	8.86
6000	1.45	1.66	2.58	5.05	8.80	15.40
10,000	1.55	1.91	3.28	7.53	14.0	26.00
20,000	1.64	2.28	4.71	12.20	26.90	52.30

* The PEG solutions contained 50 mM Tris-HCl (pH 8.0), and 0.05% Tween 20 for stable FCS measurements.

3.2. Diffusion Times (τ_D) of Compounds, Peptides, and Proteins in Viscous PEG Solutions

The effects of the various PEG solvents (PEG1000, PEG3350, PEG6000, PEG10,000, and PEG20,000 solvents) on the diffusion rates of solute molecules were measured by FCS. We measured the diffusion rates of seven solute molecules: the fluorescence dye TAMRA (Mw = 528.0), Alexa647 (1155.1), fluorescence-labeled peptide 10-mer TAMRA (1572.6), and 16-mer TAMRA (2346.6) (Figure 1), fluorescent-labeled proteins, Brd2 (16,500), secernin-1 (48,500), and IgG (150,000) (Figure 2). The Brd2 and secernin-1 samples were purified after fluorescent labelling, although the samples still contained the free-labeling reagent (29% for the Brd2 sample, and 16% for the secernin-1 sample). Therefore, the τ_D values were analyzed by the two-component analysis software of the instrument (MF20) and decomposed into two τ_D values owing to the protein and contaminating free-labeling reagent (Alexa647). For the analysis, the τ_D value of Alexa647 in each PEG solvent was measured and used. The τ_D values of each solute in the various PEG solutions were normalized by their standard $\tau_D$value (τ_{D0}), determined in FCS buffer solution without PEG. The normalized τ_D values are plotted as a function of the macroviscosity of the PEG solution in Figures 1 and 2.

Figures 1 and 2 show that, in the PEG1000 solution, the normalized τ_D values of the seven molecules almost linearly increased with increasing solvent viscosity, which mostly followed the SE equation (Equation (3)). However, the slopes of the regression curves decreased with the increasing MW of the PEGs: PEG1000 (purple) > PEG3350 (blue) > PEG6000 (green) > PEG10,000 (orange) > PEG20,000 (red). In high-MW-PEG solutions, the normalized diffusion mobility was strongly nonlinear and showed considerable deviation to faster diffusion than the SE behavior. The extent of the deviation of normalized τ_D values was more significant when the solute MW was smaller: 16-mer TAMRA (2346.6) < 10-mer TAMRA (1572.6) < Alexa647 (1155.1) < TAMRA (528.0). No clear difference was observed with Brd2 (16.5 k), secernin-1 (48.5 k), and IgG (150 k). These results showed that the extent of diffusion mobility deviations depends on both the MWs of the solute (M_{sol}) and the solvent PEG (M_{peg}). When the M_{sol} is smaller and the M_{peg} is larger, the measured τ_D becomes smaller than expected by the SE equation. Note that the normalized τ_D of the small molecular compounds did not exceed 6-fold, even in 20% PEG20,000 solution with a viscosity of 52.3 mPa·s.

Figure 3 shows the correlations between the normalized solute τ_D and the solvent viscosity, which are shown in Figures 1 and 2, in a double logarithmic plot. The correlations were linearly regressed by the least-squares method and fitted well as the following:

$$Log(\tau_D/\tau_{D0}) = LogA + \alpha Log\eta_{macro}$$

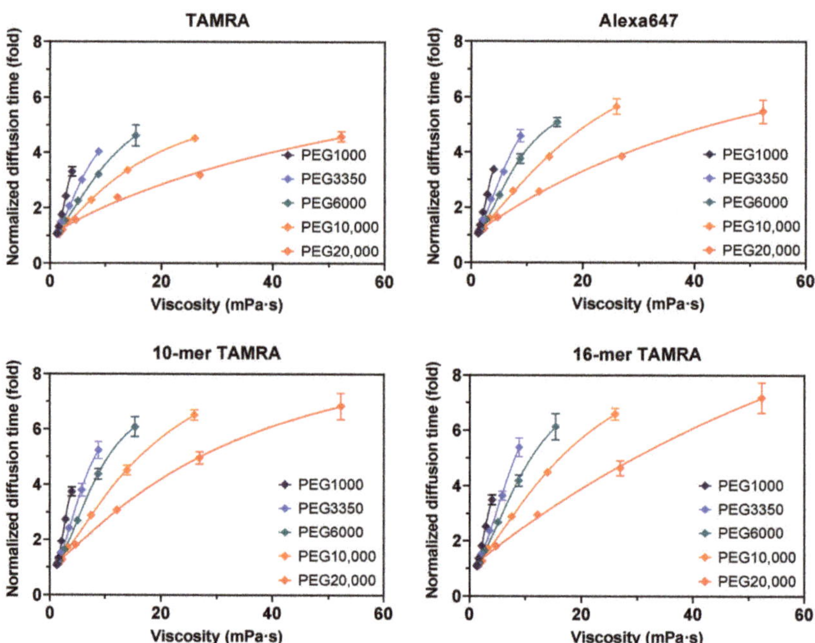

Figure 1. Measured and normalized τ_D values of TAMRA, Alexa647, 10-mer TAMRA, and 16-mer TAMRA in different PEG solvents. The τ_D values were measured and normalized by that measured in FCS buffer solution without PEG. The measurements (n = 5) were performed in duplicate. The purple, blue, green, orange, and red points are the results in the PEG1000, PEG3350, PEG6000, PEG10,000, and PEG20,000 solutions, respectively. The data were analyzed and plotted using GraphPad Prism version 8.4.2 for Windows (GraphPad Software, San Diego, CA, USA).

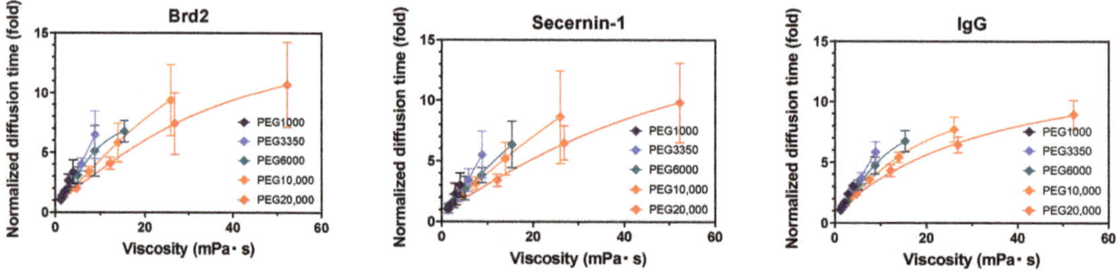

Figure 2. Normalized τ_D values of Brd2, secernin-1, and IgG in different PEG solvents. The measurements (n = 5) were performed in duplicate. The measured τ_D values were analyzed by the two-component analysis software of the instrument (MF20) and decomposed into two τ_D values owing to the protein and contaminating free-labeling reagent, then calculated τ_D values of proteins were normalized as in Figure 1.

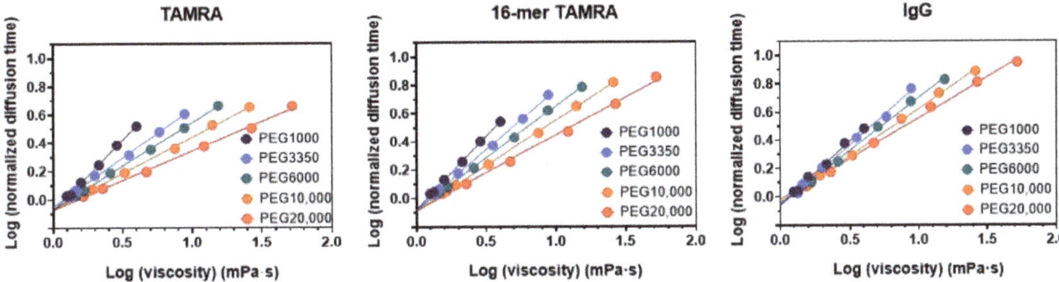

Figure 3. The measured diffusion times (τ_D) depend on macroviscosity (η_{macro}). The relationships between the normalized solute τ_Ds and solvent η_{macro}s are shown with both logarithmical axes for TAMRA, 16-mer TAMRA and IgG.

By substituting τ_D by referring to Equation (3), this equation is written as Equation (4) with an anomalous index α ($\alpha \leqq 1$).

$$Log(\tau_D/\tau_{D0}) = Log(k_2\eta_{nano}) = Logk_2 + Log\eta_{nano} = Logk_2 + \alpha Log\eta_{macro} \quad (4)$$

When SE equation holds α is 1 (Equation (3)), and α would not exceed 1, because the diffusion becomes faster than the SE relation. The estimated α values are 0.9235, 0.6628, 0.5807, 0.5088, 0.4843 and 0.393 for PEG1000, PEG3350, PEG6000, PEG8000, PEG10,000 and PEG20,000, respectively, for TAMRA. 0.9553, 0.7992, 0.6959, 0.6407, 0.6062 and 0.5049 for 16-mer TAMRA and 0.8564, 0.8257, 0.7264, 0.6631, 0.6511 and 0.5699 for IgG. The values of k_2 estimated from Equation (4) are listed in Table 2, (their average value is 0.907).

Table 2. Obtained k_4, *RSP* and k_2 values for the solutes.

Solute	M.W.	k4	*RSP*	k2 *
TAMRA	528.0	0.285	0.672	0.91
ALEXA642	1155.1	0.256	1.381	0.91
10-merTAMRA	1572.6	0.236	1.289	0.89
16-mer TAMRA	2346.6	0.213	2.437	0.90
Brd2	16,500.0	0.150	18.460	0.91
secernin-1	48,500.0	0.126	138.800	0.88
ALEXA-IgG	150,000.0	0.142	303.358	0.94

* The average of k_2 is 0.907.

3.3. *Quantitative Approximation of the Anomalous Diffusion*

It seems plausible that the extent of anomalous diffusion depends on the relative size of the solute molecule and the PEG molecule. Thus, in Figure 4, the anomalous index, α, is plotted against the ratio of the M_{sol} and the M_{peg}, M_{sol}/M_{peg}, with both logarithmic axes. We found that relations of all α and the ratio can be approximated by exponential relations as:

$$Log(\alpha) = Logk_3 + k_4 Log\left(\frac{M_{sol}}{M_{peg}}\right) = Log\left(k_3 \times \left(\frac{M_{sol}}{M_{peg}}\right)^{k_4}\right)$$

$$\alpha = k_3 \times \left(\frac{M_{sol}}{M_{peg}}\right)^{k_4} \quad (5)$$

where k_3 and k_4 are specific constants.

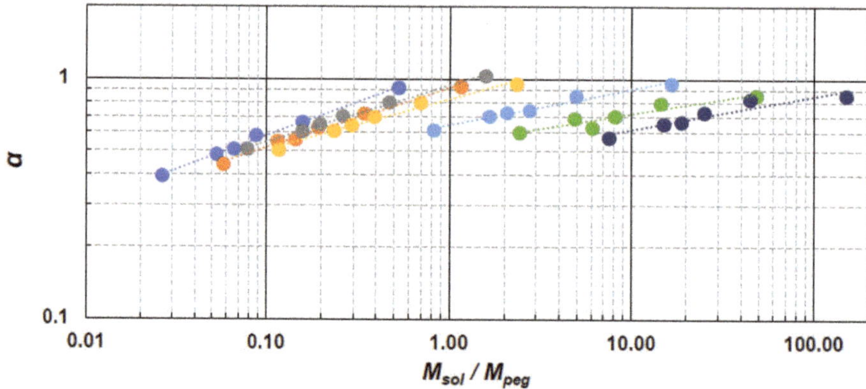

Figure 4. The anomalous index α depends on M_{sol}/M_{peg}. Blue, orange, grey, yellow, light blue, green and dark blue dots correspond to the α value of TAMRA, ALEXA, 10-mer TAMRA, 16-mer TAMRA, Brd2, sesernin-1 and ALEXA-IgG obtained in various molecular weight PEGs, that is PEG1000, PEG3350, PEG6000, PEG8000, PEG10,000 and PEG20,000. Colored dotted lines show exponential approximations for the corresponding dots.

We defined RSP as the ratio of M_{sol} and M_{peg}, where α reaches 1:

$$1 = k_3 \times RSP^{k_4} \tag{6}$$

and Equation (5) can be written as:

$$\alpha = \left(\frac{M_{sol}}{RSP \times M_{peg}}\right)^{k_4} \tag{7}$$

$$\log \alpha = k_4 \log\left(\frac{M_{sol}}{RSP \times M_{peg}}\right) = k_4 \log\left(\frac{1}{RSP}\right) + k_4 \log\left(\frac{M_{sol}}{M_{peg}}\right)$$

Then, k_4 and RSP were determined by calculating the data in Figure 4 using the method of least-squares. They are shown in Table 2.

The obtained k_4 and RSP values are plotted against M_{sol} in Figure 5 with both logarithmic axes. Both parameters can be approximated by exponentiations as follows:

$$Logk_4 = Logk_5 - k_6 LogM_{sol} \tag{8}$$

$$LogRSP = Logk_7 + k_8 logM_{sol} \tag{9}$$

where k_5, k_6, k_7 and k_8 are specific constants. The dotted lines in Figure 5 are the linear regression of the dots, showing that k_5, k_6, k_7 and k_8 are 0.6804, 0.1450, 3.819×10^{-4} and 1.1456, respectively. Then, the anomalous index, α in the Equation (7), is as follows, using the estimated values:

$$\alpha = \left(\frac{M_{sol}}{k_7 \times {M_{sol}}^{k_8} \times M_{peg}}\right)^{k_5 \times {M_{sol}}^{-k_6}} = \left(\frac{M_{sol}}{3.819 \times 10^{-4} \times {M_{sol}}^{1.1456} \times M_{peg}}\right)^{0.6804 \times {M_{sol}}^{-0.145}} \tag{10}$$

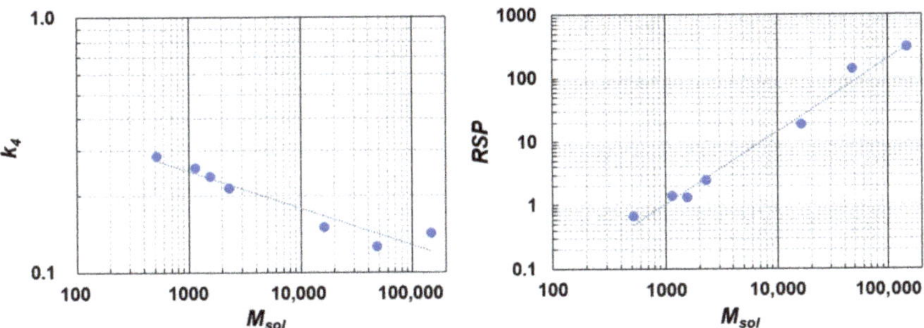

Figure 5. k_4 and RSP values plotted against M_{sol}. Both parameters can be approximated by exponentiation with M_{sol}.

Finally, using Equation (10), we are able to estimate the anomalous index α, so that τ_D of a solute molecule in an arbitrary PEG solvent can be estimated by the following equation.

$$\tau_D / \tau_{D0} = k_2 \times \eta_{macro}{}^{\alpha} \tag{11}$$

3.4. Simulation of the Diffusion Processes in the Counter Diffusion Chamber

It is important for counter-diffusion crystallization to estimate the changes over time regarding the concentrations of the protein and crystallization reagents in the counter-diffusion chamber. For the simulation program, we first estimated the macroviscosities of arbitrary MW PEGs with various concentrations. We have reported empirical equations to estimate the values as Equations (1) and (2) in the previous paper [16], so that the results of Table 1 were assigned to the equations, and the parameters were refined by the least-squares methods using the Microsoft Excel solver. The equations with the refined parameters are as follows:

$$\eta_{macro} = 1.002 \times e^{\gamma \times C_{peg}} \tag{12}$$

$$\gamma = 0.045761 \times \ln(Mpeg) - 0.2554 \tag{13}$$

where C_{peg} is the w/v% of the PEG solvent. From Equations (12) and (13), η_{macro} of PEG solvents can be calculated from their MWs and concentrations.

The changes in protein and crystallization reagents' concentrations in each area (0.5 mm length) of the counter-diffusion chamber since the start of diffusion can be estimated from each initial concentration by solving the one-dimensional diffusion partial difference equations repetitively by Microsoft Excel macro, as follows:

$$D_{pro}(x,t) = D_{pro0} \Big/ \left(0.907 \times \eta_{macro}\left(M_{peg}, C_{peg}(x,t)\right)^{\alpha(M_{pro}, M_{peg})}\right) \tag{14}$$

$$D_{sol}(x,t) = D_{sol0} \Big/ \left(0.907 \times \eta_{macro}\left(M_{peg}, C_{peg}(x,t)\right)^{\alpha(M_{sol}, M_{peg})}\right) \tag{15}$$

$$if\ Ar(x-\Delta x) \le Ar(x)\ then\ \beta_1 = \frac{Ar(x-\Delta x)}{Ar(x)}\ else\ \beta_1 = 1$$

$$if\ Ar(x+\Delta x) \le Ar(x)\ then\ \beta_2 = \frac{Ar(x+\Delta x)}{Ar(x)}\ else\ \beta_2 = 1$$

$$C_{pro}(x,t+\Delta t) = C_{pro}(x,t) + \left(\left(C_{pro}(x-\Delta x,t) - C_{pro}(x,t)\right) \times \beta_1 + \left(C_{pro}(x-\Delta x,t) - C_{pro}(x,t)\right) \times \beta_2\right) \times D_{pro}(x,t) \times \frac{\Delta t}{\Delta x^2} \tag{16}$$

$$C_{sol}(x,t+\Delta t) = C_{sol}(x,t) + \left(\left(C_{sol}(x-\Delta x,t) - C_{sol}(x,t)\right) \times \beta_1 + \left(C_{sol}(x-\Delta x,t) - C_{sol}(x,t)\right) \times \beta_2\right) \times D_{sol}(x,t) \times \frac{\Delta t}{\Delta x^2} \tag{17}$$

$$C_{peg}(x,t+\Delta t) = C_{peg}(x,t) + \left(\left(C_{peg}(x-\Delta x,t) - C_{sol}(x,t)\right) \times \beta_1 + \left(C_{peg}(x-\Delta x,t) - C_{sol}(x,t)\right) \times \beta_2\right) \times D_{peg} \times \frac{\Delta t}{\Delta x^2} \quad (18)$$

where D_{pro0}, D_{sol0} and D_{peg} are the diffusion coefficients of the protein, other solute, and PEG in water. $D_{pro}(x,t)$ and $D_{sol}(x,t)$ are the diffusion coefficients of the protein and other solute at position x in the chamber on time t. They are derived from Equations (10)–(13), using $C_{pro}(x,t)$, $C_{sol}(x,t)$ and $C_{peg}(x,t)$ which are the concentrations of the protein, other solute, and PEG at x and on t. $Ar(x)$ is the cross section of the chamber at x. β_1 and β_2 are the cross-section factors. They are determined by the ratio of the two cross sections of adjacent regions having length of Δx (in this case, 0.5 mm). The calculation step time, Δt is 20 s. According to Equations (14) and (15), the diffusion coefficients (D) of molecules (protein, other solute) in the small area at time t are calculated dividing the values in the aqueous solution (D_0) by each calculated ratio, τ_D/τ_{D0}. The τ_D/τ_{D0} of molecules was calculated according to the macroviscosity of the small areas (Equations (10) and (11)), and the value of macroviscosity is determined by concentration and MW of PEG using Equations (12) and (13). The diffusion coefficients of PEG4000, PEG20,000, NaCl, and lysozyme in water are 1.24 [29], 0.476 [29], 15.0 [30], 1.06 [31], $\times 10^{-10}$/m^{2}s^{-1}, respectively. The counter-diffusion chamber was assumed to be a gel tube with a length of 6 mm and an inner diameter of 1 mm, and a capillary with an inner diameter of 0.5 mm connected to the gel tube, and a protein sample solution was filled with 40 mm in the capillary.

Figure 6 show the results of repetitive calculation of Equations (14)–(18). Figure 6a shows the process of a case where a capillary is filled with 24 mg/mL lysozyme and a crystallization reagent is 20% PEG4000. Figure 6a1 shows the protein concentrations in various areas of the capillary after 4, 32, 64, 128, and 192 days. During this period, the lysozyme diffused out of the capillaries, and its concentration significantly decreased (Figure 6a1). Figure 6a2 shows those of PEG4000. It took time for PEG4000 to diffuse into the capillary. Figure 6a3 plots the PEG4000 concentrations of each part of the capillary on the horizontal axis and the corresponding protein concentrations on the vertical axis. The plotted curves show that the capillary could scan the protein-PEG4000 plane over time, but there is a wide unscanned area in the upper right corner of the figure. Figure 6b shows the case of 20% PEG20,000 with the same lysozyme concentration. Since the diffusion coefficient of PEG20,000 is about 1/3 of that of PEG4000, it took more time to diffuse into the capillary (Figure 6b2). During this period, the lysozyme diffused out of the capillaries, and its concentration significantly decreased (Figure 6b1). As a result, the scanned area of the protein-PEG plane becomes narrower (Figure 6b3). These results show that the self-searching mechanism for the optimal crystallization scenario would not work well when high-MW PEG is used as a precipitant in the counter-diffusion method.

These results indicated that viscous PEG solvents are not good precipitants for counter-diffusion crystallization. On the other hand, high-MW PEG is expected to improve the quality of crystals due to its dehydration effect [32]. Thus, PEG is a reagent that we would like to try.

Regarding the nucleation, the probability is explained by the following equation [33]:

$$\frac{\partial N}{\partial t} = V \times \frac{const}{\eta} \times \exp\left(-\frac{16\pi v^2 \gamma^3}{3(kT)^3 [lnS]^2}\right) \quad (19)$$

where $\partial N/\partial t$, S, γ, and v, are the nucleation probability, supersaturation, surface energy, and volume of the crystal. V, η and *const* are the volume of the solution, viscosity and the constant, which are related to the attachment kinetics of growth units. This depends on the molecular charge, the molecular volume, and the density of the solution. Therefore, it is expected that the nucleation probability will decrease in highly viscous solution.

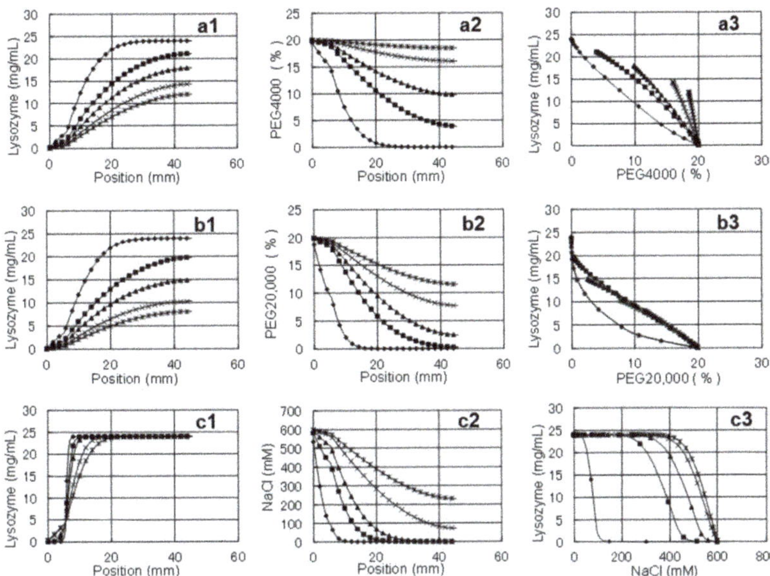

Figure 6. Calculated results of the diffusion processes in the crystallization chamber with 1D simulation. The results of repetitive calculation of Equations (14)–(18) are shown. The crystallization chamber is assumed to be a 6 mm length and 1 mm bore gel tube and 0.5 mm bore glass capillary with a 40 mm length protein solution; (**a**) 24 mg/mL of lysozyme is loaded into the capillary and 20% of PEG4000 is applied as the precipitant for the reservoir solution; (**b**) 24 mg/mL of lysozyme and 20% of PEG20,000; (**c**) 24 mg/mL of lysozyme with 20% of PEG20,000 is loaded into the capillary, 20% of PEG20,000 is presoaked in the gel tube and 20% of PEG20,000 and 600 mM of NaCl is applied as the precipitant for the reservoir solution. The protein concentrations along the chamber are shown in (**a1**,**b1**,**c1**). The abscissa shows the position along the chamber from the open end of the gel tube. The concentrations of the precipitant along the chamber are shown in (**a2**,**b2**,**c2**). The concentration relations of the precipitant and the protein along the chamber are plotted in (**a3**,**b3**,**c3**). For (**a**,**b**), ◆, ■, ▲, ×, Ж show the values of 4, 32, 64, 128 and 192 days after the starting of the diffusion, respectively. For (**c**), they show the values of 0.2, 1, 2, 8 and 16 days, respectively. The values are plotted every 2 mm along the chamber.

We reported the nucleation probability in Table 1 of our previous report [34]. The lysozyme was crystallized by batch method with a 50 mM acetate buffer of pH 4.5 and various amounts of PEG4000 and NaCl. The concentrations of NaCl tested were 300, 500, 800 and 1000 mM, and of PEG4000, they were 0, 5, 10, 15, 20 and 25%. The nucleation probability decreased in the higher concentration PEG solvents, almost inversely proportional to the τ_D/τ_{D0} values. The result also suggested the possibility of viscous PEG solvents to obtain fewer but larger crystals.

PEG and small molecules, such as salts, synergistically act on protein crystallization [35]. Therefore, in Figure 6c, we have simulated the following case: a protein sample is pre-mixed with 20% PEG20,000 in advance and crystallization begins upon diffusing 600 mM NaCl contained in 20% PEG20,000 solvent into the capillary. The PEG20,000 concentration was pre-uniform in the counter-diffusion chamber. The concentrations were plotted 0.2, 1, 2, 8 and 16 days after the start of diffusion. The results (Figure 6c3) showed that, after 16 days, most of the protein-NaCl plane was scanned. From these results, it was strongly suggested that crystallization using the counter-diffusion method with high-MW PEG is both possible and promising.

3.5. Quality of Crystals Grown in Viscous PEG Solvents Using Counter-Diffusion Systems

To confirm the crystallization performance obtained by diffusing a small-molecule precipitant into the capillary, where protein samples were pre-mixed with viscous PEG solution in advance, lysozyme crystals were grown in PEG4000, PEG10,000, and PEG20,000 solutions using the counter-diffusion method with gel-tube parts. A lysosome solution with 20 mg protein/mL in each 20% PEG solution containing 50 mM acetate buffer (pH 4.5) was filled in the capillaries and placed in reservoir solutions containing the corresponding 20% PEG, 50 mM acetate buffer (pH 4.5), and 600 mM NaCl at 293 K. After incubation for 20 days at 293 K, crystals were observed in all the capillaries and, after incubation for 30 days, the crystals were harvested. In the X-ray diffraction experiment, three or four crystals grown in each PEG solution were used and the datasets were collected. The crystal-to-detector distance was fixed to the maximum resolution of 1.12 Å. For the crystal from the PEG20,000 solution, an additional dataset with the distance fixed to 1.02 Å was also collected from the same crystal. The dataset from the crystal with the lowest *B* factor of the Wilson plot was selected, and the statistics are listed in Table 3.

Table 3. Data collection and scaling statistics. The values in parentheses are for the highest resolution shells.

	Crystal ID			
	4834BK1	**4835M**	**4836M**	**4836M ***
Major precipitant	PEG4000	PEG10,000	PEG20,000	PEG20,000
Wavelength (Å)	0.8266	0.8266	0.8266	0.7514
Maximum resolution (Å)	1.12	1.12	1.12	1.02
Oscillation range (°)	0.1	0.1	0.1	0.1
Number of images	3600	3600	3600	3600
X-ray exposure time per frame (s)	0.01	0.01	0.01	0.02
Space group	$P4_32_12$	$P4_32_12$	$P4_32_12$	$P4_32_12$
Unit-cell parameters (Å)	a = 77.27, b = 77.27, c = 37.55	a = 77.27, b = 77.27, c = 37.83	a = 77.15, b = 77.15, c = 37.62	a = 77.14, b = 77.14, c = 37.62
Resolution range (Å)	54.64–1.12 (1.14–1.12)	54.64–1.12 (1.14–1.12)	54.55–1.12 (1.14–1.12)	54.55–1.02 (1.04–1.02)
No. of observed reflections	1,033,331 (52,069)	1,058,437 (53,038)	1,040,904 (47,802)	1,392,055 (63,330)
No. of unique reflections	43,853 (2159)	44,169 (2164)	43,886 (2064)	58,018 (2687)
Multiplicity	23.6 (24.1)	24.0 (24.5)	23.7 (23.2)	24.0 (23.6)
Completeness (%)	100 (100)	100 (100)	99.8 (96.4)	99.8 (95.5)
$<I>/<\sigma(I)>$	18.2 (5.8)	27.3 (5.5)	24.6 (7.8)	20.7 (4.1)
R_{merge}	10.0 (57.3)	5.5 (56.7)	8.1 (38.9)	8.3 (78.5)
Wilson *B* factor (Å^2)	9.2	10.1	8.5	8.4
Overall *B* factor from relative Wilson plot (Å^2)	−0.34	−0.8	0	0
Mosaicity	0.42 ± 0.11	0.31 ± 0.13	0.17 ± 0.06	0.18 ± 0.06

* The dataset was collected with the proper crystal-to-detector distance, and observed the maximum resolution of 1.02 Å.

The parameters that are usually used to evaluate the protein-crystal quality are the maximum resolution, R_{merge}, $<I>/<\sigma(I)>$, *B* factor of the Wilson plot, and mosaicity [36,37]. Among these parameters, the maximum resolution, R_{merge}, and $<I>/<\sigma(I)>$ depend on the experimental diffraction conditions, such as the size of the crystals and intensity of the X-ray. Conversely, the overall *B* factor obtained from the relative Wilson plot is independent of the experimental diffraction conditions [36]. The Wilson *B* factor of the crystal grown in the PEG20,000 solution was the smallest, and the overall *B* factor largest, indicating that the quality of the crystal grown in the solution was better than the others. Regarding the mosaicity, the average value (0.17) and deviation (0.06) were the lowest for the crystal grown in the PEG20,000 solution. The mosaicities of all of the frames are plotted in Figure 7. They showed that the anisotropy was smaller for the crystal grown from the PEG20,000 solution.

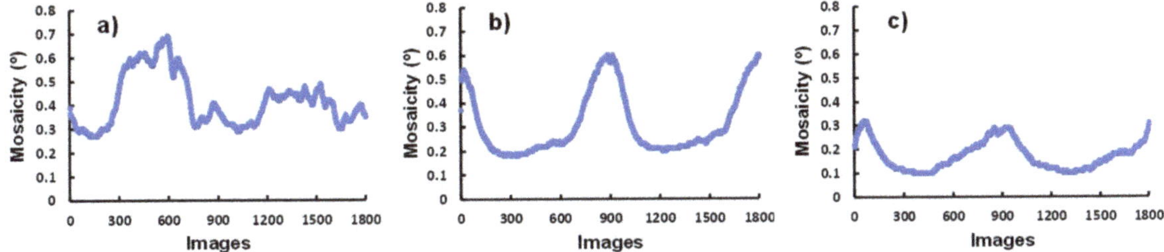

Figure 7. Mosaicities are plotted against the continuous frames: (**a**) 4834BK1 from PEG4000, (**b**) 4835M from PEG10,000 and (**c**) 4836M from PEG20,000.

In this study, lysozyme crystals with good diffractive quality were obtained in the viscous PEG solution using the counter-diffusion method. Their maximum resolutions were comparable to or better than the already registered data for a tetragonal lysozyme crystal in the PDB (for example, PDB ID: 4hp0, 5kxz, 6g8a), except for the crystal grown in space (1iee). Among the three MW PEGs, the best crystals were grown in PEG20,000. There are three possible reasons for this. (1) The density-driven flow around the growing crystals is reduced when the viscosity of the solution is higher. As a result, the highly ordered attachment of the protein molecules to the crystal surface suffers less from the flow [38,39]. (2) In a highly viscous solution, the reduction in the density-driven flow and the decrease of diffusive migration of the protein molecules enhances the formation of the protein depletion zone around the growing crystal [4]. As a result, the protein concentration on the surface of the growing crystal decreases, allowing for good crystals to be grown under lower supersaturation. (3) The excluded volume effect is stronger in high-MW-PEG solutions [40]. This generates a much stronger force to drive the protein molecules onto the crystal surface. The stronger macroscopic force on the surface of the crystal moves the protein molecules closer together inside the crystals and the molecular arrangement in the crystals becomes better. The mechanisms of growing good crystals in viscous high-MW-PEG solutions will be revealed by further investigations.

4. Conclusions

The diffusion mobility of solutes in viscous PEG solvents shows a considerable deviation, which is faster than the expected diffusion rate determined from the macroviscosity. This behavior was quantitatively described in this paper using an approximation model. The model enabled us to describe the mass transportation of molecules during the counter-diffusion crystallization processes. Figure 6a,b clarified that the self-searching mechanism for the optimal crystallization scenario would not work well when high-MW PEG is used as a precipitant in the counter-diffusion method.

However, crystallization in the counter-diffusion chamber containing a uniform concentration of PEG20,000 in advance works well. The greater extent of anomalous diffusion in a high-MW-PEG solution was observed with small-molecule solutes as salt precipitants, so that without a large delay in the crystallization period, the protein solution pre-mixed with viscous high-MW-PEG solutions could be crystalized using salt precipitants in the counter-diffusion crystallization method. Figure 6c shows that the self-searching mechanism worked well under this condition, and most of the protein-NaCl plane could be scanned after 16 days. We further confirmed the possibility of crystallization experiments with lysozyme solution pre-mixed with viscous high-MW PEG.

Author Contributions: Conceptualization, H.T. and A.T.; sample preparation and FCS experiments, R.U., K.T. and A.T.; crystallization experiment, A.T.; validation, A.T.; X-ray structural analysis, S.T. and H.T.; writing—original draft preparation, A.T.; writing—review and editing, A.T. and

H.T.; visualization, A.T. and H.T.; supervision, A.T. and H.T.; project administration, A.T.; funding acquisition, A.T. All authors have read and agreed to the published version of the manuscript.

Funding: This work was supported in part by the Development of Fundamental Technology for Protein Analyses and the Target Protein Research Programs from the Ministry of Education, Culture, Sports, Science and Technology of Japan.

Institutional Review Board Statement: Not applicable.

Informed Consent Statement: Not applicable.

Data Availability Statement: Not applicable.

Acknowledgments: We are grateful to Shigeyuki Yokoyama and Mikako Shirouzu (RIKEN) for their encouraging support.

Conflicts of Interest: The authors declare no conflict of interest.

References

1. McPherson, A.; Gavira, J.A. Introduction to protein crystallization. *Acta Crystallogr. F Struct. Biol. Commun.* **2014**, *70*, 2–20. [CrossRef] [PubMed]
2. McPherson, A.; Cudney, B. Optimization of crystallization conditions for biological macromolecules. *Acta Crystallogr. F Struct. Biol. Commun.* **2014**, *70*, 1445–1467. [CrossRef] [PubMed]
3. Holcomb, J.; Spellmon, N.; Zhang, Y.; Doughan, M.; Li, C.; Yang, Z. Protein crystallization: Eluding the bottleneck of X-ray crystallography. *AIMS Biophys.* **2017**, *4*, 557–575. [CrossRef] [PubMed]
4. Garcia-Ruiza, J.M.; Novella, M.L.; Moreno, R.; Gavira, J.A. Agarose as crystallization media for proteins I: Transport processes. *J. Cryst. Growth* **2001**, *232*, 165–172. [CrossRef]
5. Gavira, J.A.; Cera-Manjarres, A.; Ortiz, K.; Mendez, J.; Jimenez-Torres, J.A.; Patino-Lopez, L.D.; Torres-Lugo, M. Use of Cross-Linked Poly(ethylene glycol)-Based Hydrogels for Protein Crystallization. *Cryst. Growth Des.* **2014**, *14*, 3239–3248. [CrossRef]
6. Weber, P.; Pissis, C.; Navaza, R.; Mechaly, A.E.; Saul, F.; Alzari, P.M.; Haouz, A. High-Throughput Crystallization Pipeline at the Crystallography Core Facility of the Institut Pasteur. *Molecules* **2019**, *24*, 4451. [CrossRef]
7. Lieske, J.; Cerv, M.; Kreida, S.; Komadina, D.; Fischer, J.; Barthelmess, M.; Fischer, P.; Pakendorf, T.; Yefanov, O.; Mariani, V.; et al. On-chip crystallization for serial crystallography experiments and on-chip ligand-binding studies. *IUCrJ* **2019**, *6*, 714–728. [CrossRef]
8. De Wijn, R.; Hennig, O.; Roche, J.; Engilberge, S.; Rollet, K.; Fernandez-Millan, P.; Brillet, K.; Betat, H.; Morl, M.; Roussel, A.; et al. A simple and versatile microfluidic device for efficient biomacromolecule crystallization and structural analysis by serial crystallography. *IUCrJ* **2019**, *6*, 454–464. [CrossRef]
9. Garcia-Ruiz, J.M.; Moreno, A. Investigations on protein crystal growth by the gel acupuncture method. *Acta Crystallogr. D Struct. Biol. Commun.* **1994**, *D50*, 484–490. [CrossRef]
10. Hashizume, Y.; Inaka, K.; Furubayashi, N.; Kamo, M.; Takahashi, S.; Tanaka, H. Methods for Obtaining Better Diffractive Protein Crystals: From Sample Evaluation to Space Crystallization. *Crystals* **2020**, *10*, 78. [CrossRef]
11. Otalora, F.; Garcia-Ruiz, J.M. Computer model of the diffusion/ reaction interplay in the gel acupuncture method. *J. Cryst. Growth* **1996**, *169*, 361–367. [CrossRef]
12. Garcia-Ruiz, J.M.; Otalora, F.; Garcia-Caballero, A. The role of mass transport in protein crystallization. *Acta Crystallogr. F Struct. Biol. Commun.* **2016**, *72*, 96–104. [CrossRef]
13. Tanaka, H.; Inaka, K.; Sugiyama, S.; Takahashi, S.; Sano, S.; Sato, M.; Yoshitomi, S. A simplified counter diffusion method combined with a 1D simulation program for optimizing crystallization conditions. *J. Synchrotron Radiat.* **2004**, *11*, 45–48. [CrossRef]
14. Liu, J.; Wu, S.; Cao, D.; Zhang, L. Effects of pressure on structure and dynamics of model elastomers: A molecular dynamics study. *J. Chem. Phys.* **2008**, *129*, 154905. [CrossRef]
15. Li, C.; Wang, Y.; Pielak, G.J. Translational and rotational diffusion of a small globular protein under crowded conditions. *J. Phys. Chem. B* **2009**, *113*, 13390–13392. [CrossRef]
16. Tanaka, H.; Takahashi, S.; Yamanaka, M.; Fukuyama, S.; Sano, S.; Motohara, M.; Kobayashi, T.; Yoshitomi, S.; Tanaka, T. Diffusion Coefficient of the Protein in Various Crystallization Solutions: The Key to Growing High-quality Crystals in Space. *Microgravity Sci. Technol.* **2006**, *18*, 91–94. [CrossRef]
17. Holyst, R.B.A.; Szymanski, J.; Wilk, A.; Patkowski, A.; Gapinski, J.; Zywocinski, A.; Kalwarczyk, T.; Kalwarczyk, E.; Tabaka, M. Scaling form of viscosity at all length-scales in poly(ethylene glycol) solutions studied by fluorescence correlation spectroscopy and capillary electrophoresis. *Phys. Chem. Chem. Phys.* **2009**, *11*, 9025–9032. [CrossRef]
18. Haustein, E.; Schwille, P. Fluorescence correlation spectroscopy: Novel variations of an established technique. *Annu. Rev. Biophys. Biomol. Struct.* **2007**, *36*, 151–169. [CrossRef]
19. Kigawa, T.; Yabuki, T.; Yoshida, Y.; Tsutsui, M.; Ito, Y.; Shibata, T.; Yokoyama, S. Cell-free production and stable-isotope labeling of milligram quantities of proteins. *FEBS Lett.* **1999**, *442*, 15–19. [CrossRef]

20. Kigawa, T. Cell-free protein production system with the E. coli crude extract for determination of protein folds. *Methods Mol. Biol.* **2010**, *607*, 101–111. [CrossRef]
21. Kigawa, T.; Yabuki, T.; Matsuda, N.; Matsuda, T.; Nakajima, R.; Tanaka, A.; Yokoyama, S. Preparation of Escherichia coli cell extract for highly productive cell-free protein expression. *J. Struct. Funct. Genom.* **2004**, *5*, 63–68. [CrossRef] [PubMed]
22. Nakamura, Y.; Umehara, T.; Nakano, K.; Jang, M.K.; Shirouzu, M.; Morita, S.; Uda-Tochio, H.; Hamana, H.; Terada, T.; Adachi, N.; et al. Crystal structure of the human BRD2 bromodomain: Insights into dimerization and recognition of acetylated histone H4. *J. Biol. Chem.* **2007**, *282*, 4193–4201. [CrossRef] [PubMed]
23. Tsuganezawa, K.; Watanabe, H.; Parker, L.; Yuki, H.; Taruya, S.; Nakagawa, Y.; Kamei, D.; Mori, M.; Ogawa, N.; Tomabechi, Y.; et al. A novel Pim-1 kinase inhibitor targeting residues that bind the substrate peptide. *J. Mol. Biol.* **2012**, *417*, 240–252. [CrossRef]
24. Tsuganezawa, K.; Shinohara, Y.; Ogawa, N.; Tsuboi, S.; Okada, N.; Mori, M.; Yokoyama, S.; Noda, N.N.; Inagaki, F.; Ohsumi, Y.; et al. Two-colored fluorescence correlation spectroscopy screening for LC3-P62 interaction inhibitors. *J. Biomol. Screen.* **2013**, *18*, 1103–1109. [CrossRef]
25. Yamanaka, M.; Inaka, K.; Furubayashi, N.; Matsushima, M.; Takahashi, S.; Tanaka, H.; Sano, S.; Sato, M.; Kobayashi, T.; Tanaka, T. Optimization of salt concentration in PEG-based crystallization solutions. *J. Synchrotron Radiat.* **2011**, *18*, 84–87. [CrossRef] [PubMed]
26. Battye, T.G.; Kontogiannis, L.; Johnson, O.; Powell, H.R.; Leslie, A.G. iMOSFLM: A new graphical interface for diffraction-image processing with MOSFLM. *Acta Crystallogr. D Biol. Crystallogr.* **2011**, *67*, 271–281. [CrossRef]
27. Evans, P.R.; Murshudov, G.N. How good are my data and what is the resolution? *Acta Crystallogr. D Biol. Crystallogr.* **2013**, *69*, 1204–1214. [CrossRef]
28. Winn, M.D.; Ballard, C.C.; Cowtan, K.D.; Dodson, E.J.; Emsley, P.; Evans, P.R.; Keegan, R.M.; Krissinel, E.B.; Leslie, A.G.; McCoy, A.; et al. Overview of the CCP4 suite and current developments. *Acta Crystallogr. D Biol. Crystallogr.* **2011**, *67*, 235–242. [CrossRef]
29. Luo, Z.; Zhang, G. Scaling for sedimentation and diffusion of poly(ethylene glycol) in water. *J. Phys. Chem. B* **2009**, *113*, 12462–12465. [CrossRef]
30. Vitagliano, V.; Lyons, A. Diffusion coefficients for aqueous solutions of sodium chloride and barium Chroride. *J. Am. Chem. Soc.* **1955**, *78*, 1549–1552. [CrossRef]
31. Dubin, S.B.; Clark, N.A.; Benedek, G.B. Measurement of the rotational diffusion coefficient of lysozyme by depolarized light scattering: Configration of lysozyme in solution. *J. Chem. Phys.* **1971**, *54*, 5158. [CrossRef]
32. Heras, B.; Edeling, M.A.; Byriel, K.A.; Jones, A.; Raina, S.; Martin, J.L. Dehydration converts DsbG crystal diffraction from low to high resolution. *Structure* **2003**, *11*, 139–145. [CrossRef]
33. Manuel Garcia-Ruiz, J. Nucleation of protein crystals. *J. Struct. Biol.* **2003**, *142*, 22–31. [CrossRef]
34. Tanigawa, N.T.S.; Yan, B.; Kamo, M.; Furubayashi, N.; Kubota, K.; Inaka, K.; Tanaka, H. Novel Device and Strategy for Growing Large, High-Quality Protein Crystals by Controlling Crystallization Conditions. *Crystals* **2021**, *11*, 1311. [CrossRef]
35. Majeed, S.; Ofek, G.; Belachew, A.; Huang, C.C.; Zhou, T.; Kwong, P.D. Enhancing protein crystallization through precipitant synergy. *Structure* **2003**, *11*, 1061–1070. [CrossRef]
36. Arai, S.; Chatake, T.; Suzuki, N.; Mizuno, H.; Niimura, N. More rapid evaluation of biomacromolecular crystals for diffraction experiments. *Acta Crystallogr. D Biol. Crystallogr.* **2004**, *60*, 1032–1039. [CrossRef]
37. Wuttke, J. Multiple Bragg reflection by a thick mosaic crystal. *Acta Crystallogr. A Found. Adv.* **2014**, *70*, 429–440. [CrossRef]
38. Vekilov, P.G.; Alexander, J.I.; Rosenberger, F. Nonlinear response of layer growth dynamics in the mixed kinetics-bulk-transport regime. *Phys. Rev. E Stat. Phys. Plasmas Fluids Relat. Interdiscip. Top.* **1996**, *54*, 6650–6660. [CrossRef]
39. Gliko, O.; Booth, N.A.; Vekilov, P.G. Step bunching in a diffusion-controlled system: Phase-shifting interferometry investigation of ferritin. *Acta Crystallogr. D Biol. Crystallogr.* **2002**, *58*, 1622–1627. [CrossRef]
40. Bhat, R.; Timasheff, S.N. Steric exclusion is the principal source of the preferential hydration of proteins in the presence of polyethylene glycols. *Protein Sci.* **1992**, *1*, 1133–1143. [CrossRef]

Article

The Effect of Controlled Mixing on ROY Polymorphism

Margot Van Nerom [1,*], Pierre Gelin [2], Mehrnaz Hashemiesfahan [2,3], Wim De Malsche [2], James F. Lutsko [4], Dominique Maes [1,*] and Quentin Galand [1]

1 Structural Biology Brussels, Vrije Universiteit Brussel, 1050 Brussels, Belgium; quentin.galand@vub.be
2 Department of Chemical Engineering, Vrije Universiteit Brussel, 1050 Brussels, Belgium; pierre.gelin@vub.be (P.G.); mehrnaz.hashemiesfahan@vub.be (M.H.); wim.de.malsche@vub.be (W.D.M.)
3 Mesoscale Chemical Systems, University of Twente, 7522 NB Enschede, The Netherlands
4 Center for Nonlinear Phenomena and Complex Systems, Université Libre de Bruxelles, 1050 Brussels, Belgium; jim.lutsko@ulb.be
* Correspondence: margot.van.nerom@vub.be (M.V.N.); dominique.maes@vub.be (D.M.)

Abstract: We report the investigation of various experimental conditions and their influence on polymorphism of 5-methyl-2-[(2-nitrophenyl)amino]-3-thiophenecarbonitrile, commonly known as ROY. These conditions include an in-house-developed microfluidic chip with controlled mixing of parallel flows. We observed that different ROY concentrations and different solvent to antisolvent ratios naturally favored different polymorphs. Nonetheless, identical samples prepared with different mixing methods, such as rotation and magnetic stirring, consistently led to the formation of different polymorphs. A fourth parameter, namely the confinement of the sample, was also considered. Untangling all those parameters and their influences on polymorphism called for an experimental setup allowing all four to be controlled accurately. To that end, we developed a novel customized microfluidic setup allowing reproducible and controlled mixing conditions. Two parallel flows of antisolvent and ROY dissolved in solvent were infused into a transparent microchannel. Next, slow and progressive mixing could be obtained by molecular diffusion. Additionally, the microfluidic chip was equipped with a piezoceramic element, allowing the implementation of various mixing rates by acoustic mixing. With this device, we demonstrated the importance of parameters other than concentration on the polymorphism of ROY.

Keywords: ROY; polymorphism; microfluidics; acoustic mixing; diffusive mixing

Citation: Van Nerom, M.; Gelin, P.; Hashemiesfahan, M.; De Malsche, W.; Lutsko, J.F.; Maes, D.; Galand, Q. The Effect of Controlled Mixing on ROY Polymorphism. *Crystals* **2022**, *12*, 577. https://doi.org/10.3390/cryst12050577

Academic Editor: Abel Moreno

Received: 22 March 2022
Accepted: 14 April 2022
Published: 20 April 2022

1. Introduction

Polymorphism in crystallography was first defined by Eilhardt Mitscherlich in the 19th century and refers to the property of some chemical compositions to exist in multiple crystalline forms [1]. These different crystal structures are often a result of different molecular shapes caused by different torsion angles. Indeed, free rotation about single bonds within a molecule allows for several arrangements with potential energy minima and are therefore considered stable conformations. Consequently, the distinct molecular shapes result in different packing configurations. This particular form of polymorphism is termed conformational polymorphism [2]. However, different conformations exhibit different properties, such as stability, dissolvability, physiological activity, and/or bioavailability. This explains the importance of studies on crystal polymorphism in the pharmaceutical field. In drug development, the main objective is to achieve specific properties, and contamination with an undesired polymorph would be detrimental [3–6].

A model compound for studying polymorphism is 5-methyl-2-[(2-nitrophenyl)amino]-3-thiophenecarbonitrile or ROY because of its ability to form at least 11 polymorphs, of which 6 are stable at room temperature. The latter are yellow prisms (Ys), orange needles (ONs), orange plates (OPs), red prisms (Rs), yellow needles (YNs), and orange-red plates (ORPs) and are formed by increasing the antisolvent-to-solvent ratio, which are water and acetone, respectively. The less stable and consequently less characterized forms are red plates (RPL), Y04, R05, YT04, and PO13. Red plates crystallize from vapor on succinic acid, Y04 and R05 can be obtained from a melt crystallization, and YT04 is a transformed form of Y04. PO13 is a supercooled melt form of YNs [7]. This abundance of polymorphs can be explained by the presence of three torsion angles and their different conformations. Moreover, the polymorphs are easily detected according to their color and shape, an aspect represented in the acronym Red Orange Yellow (ROY) [5–8].

Creating a controlled mixing regime that allows the selection of one particular polymorph with its specific properties is relevant for multiple industrial disciplines [3,9]. However, the crystallization of polymorphic compounds is a complex process and depends on many parameters, such as temperature, the solvent-to-antisolvent ratio, antisolvent properties, concentration, etc. As a first step toward a controlled and constant mixing environment, this work reports on the implementation of a microfluidic chip to achieve different controlled mixing regimes [10].

2. Materials and Methods

2.1. Chemicals

5-methyl-2-[(2-nitrophenyl)amino]-3-thiophenecarbonitrile (ROY) was purchased from TRC Canada® (Toronto, Canada) and consisted of a mixture of polymorphs with sizes ranging from about 1 to 200 µm (Figure 1). For stock solutions, synthesis-grade acetone from Acros Organics® (Geel, Belgium) with a purity of 99.9% and milliQ water were used. Stock solutions (starting solutions with ROY and antisolvent solutions without ROY) were prepared with a Sartorius CPA224S scale (Goettingen, Germany) with an accuracy of 0.1 mg. In the following, the concentrations of the samples are mentioned according to "acetone (Vol%)/Water (Vol%)/ROY (mg/mL)". The volumes of all prepared stock solutions were selected to be sufficiently large (>20 mL) so that the concentrations of all samples were reported with an accuracy of volume fractions below 0.01 Vol% and accuracy of ROY concentrations below 0.1 mg/mL.

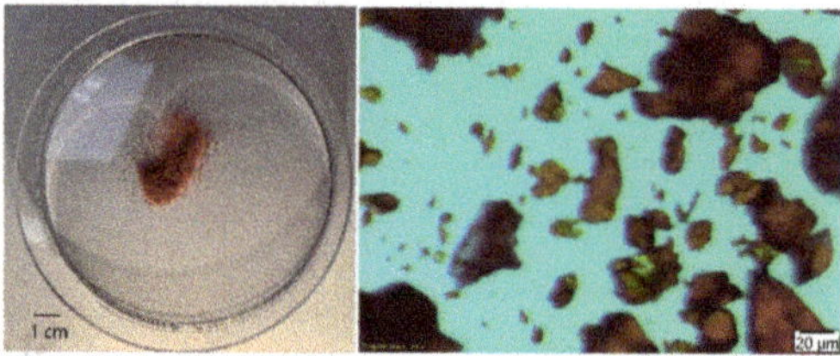

Figure 1. ROY used in this study produced a mixture of polymorphs with sizes ranging from 1 to 200 µm.

2.2. Crystallization in Bulk Experiments

Antisolvent solution was added to and mixed with starting solution up to a total volume of 1.0 mL in 1.5 mL transparent vials at room temperature (RT). The respective volumes of the infused solutions were adjusted with KD Scientific® model 100 syringe pumps (Holliston, MA, USA) using Hamilton® Gastight® Luer Lock syringes (Allston, MA, USA). Connections between the syringes and the vials were realized with Idex connectors and 450 µm inner diameter (i.d.) and 670 µm outer diameter (o.d.) glass capillaries. The flow rate was set for each sample such that the total infusion time per sample was 10 min. During infusion, mixing was performed by rotating an 8 mm magnet at 100 RPM unless otherwise mentioned, or by shaking the vial with an Eppendorf thermomixer comfort at 500 RPM unless otherwise mentioned. Evaporation of the sample was prevented by sealing the glass vials (Figure 2). Each result documented in the following is based on at least three experimental runs unless otherwise mentioned.

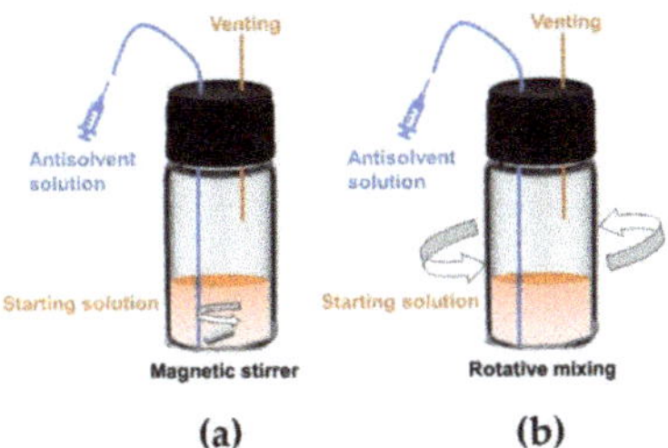

Figure 2. Setup of crystallization in bulk experiments: (**a**) sample is mixed by a magnetic stirrer during 10 min infusion; (**b**) mixed by rotation during 10 min infusion.

2.3. Crystzallization in Confinement

To evaluate the possible influence of volume on the crystallization process of ROY, mixing was first performed as mentioned in the previous Section 2.2. Immediately after mixing and before crystallization occurred, the final solution was administered to capillaries with an i.d. of 100 μm, 200 μm, or 400 μm (Borosilicate glass round tubes from CM Scientific® (Silsden, UK)). The capillaries were then sealed with melted wax (Figure 3).

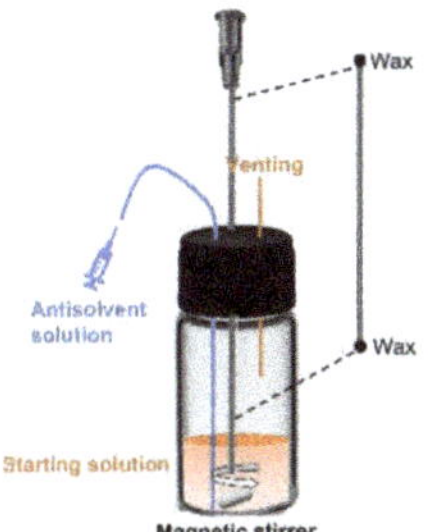

Figure 3. Confinement experiment in which sample is administered to capillaries before nucleation. Capillary is sealed afterwards with wax.

2.4. Polymorph Identification

Samples were analyzed with an Olympus® SZX16 microscope (Tokyo, Japan). The combination of the magnification of the microscope and the objective of the Olympus® SC100 color camera (Tokyo, Japan) resulted in a magnification of 23×. Different ROY polymorphs are so distinct in structure and color that optical identification was an easy and suitable method for this study. Moreover, experiments were repeated at least three times to fully characterize the observed sample.

2.5. Microfluidic Setup

All experiments were performed in microfluidic chips. The microchannel was etched in a silicon wafer, and the top and bottom of the channel were sealed with borosilicate glass. The internal section (width × depth) of the channels used in this study was 0.375 mm × 0.525 mm, and the channel length was 20 mm (Figure 4c,d). The channels were equipped with three inlets and outlets for infusion and evacuation of liquids and sealed by bonding 200 μm internal diameter Polymicro Technologies® (Phoenix, AZ, USA) glass capillaries with dual-cure epoxy sealant (Figure 4b). All connections were realized with IDEX connectors. Starting solution and antisolvent solution were simultaneously infused, and accurate flow rate controls were performed using Fluigent® pressure flow controllers (Lowell, MA, USA) (Figure 4a).

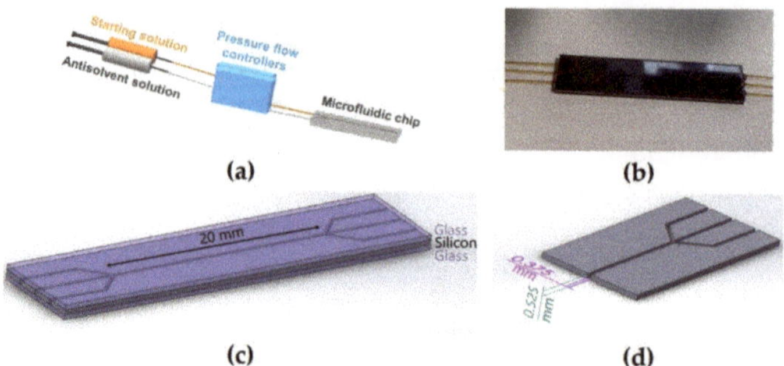

Figure 4. Microfluidic setup: (**a**) Simultaneous infusion of starting solution and antisolvent solution with accurate flow controls were performed with pressure flow controllers; (**b**) inlets and outlets were sealed with glass capillary tubes; (**c**) channel length of the chips was 20 mm; (**d**) internal section (width × depth) of the chips was 375 μm × 525 μm.

In-chip liquid mixing was performed by inducing acoustic streaming, for which a piezo-ceramic actuator (15 mm × 20 mm × 1 mm, APC international, Mackeyville, PA USA) with an eigenfrequency of around 2.0 MHz was placed at the back of the chip (Figure 5a). A Tektronix AFG1062 function generator was used to apply a sinusoidal voltage to the piezo element and an RF power amplifier (210 L, Electronics & Innovations, Rochester, NY USA) amplified the applied voltage with a maximal total output power of 10 W. To prevent the heating of the sample with the supplied acoustic energy, active temperature control was implemented: the sample was thermostabilized by a Peltier element driven by a PID controller with feedback from a thermistor incorporated in an aluminum block close to the working volume. A 120 W Peltier element was used, and a constant temperature water loop was applied at the backside of the thermoelectric element to improve the performance of the system (Figure 5b).

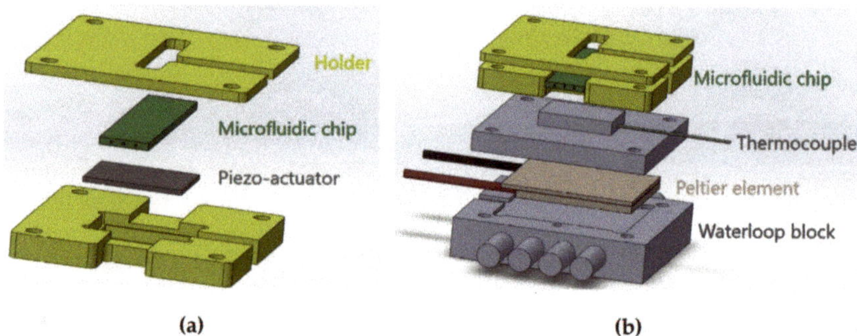

Figure 5. In-chip liquid mixing: (**a**) acoustic mixing of the liquid was obtained by vibrating the window of the chip with a piezo-ceramic actuator; (**b**) the cell system incorporated a PID temperature control system.

The overall setup was installed on an optical bench and imaged with a color HD camera through a 4× magnification objective. The microchannel was fixed on a translation stage, and the entire length of the channel could be placed in the field of view of the camera when searching for polymorphs. An overview of the experimental setup is pictured in Figure 6.

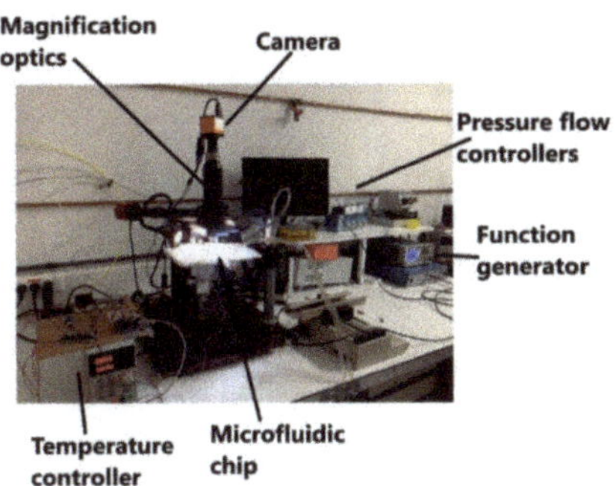

Figure 6. Overview of microfluidic experimental setup, including the PID temperature control system and the pressure flow controllers.

3. Results

3.1. Supersaturation Protocol Influences ROY Polymorphism

The most important parameters of crystallization of any compound are the concentration and the solvent-to-antisolvent ratio. ROY polymorphism has proven to be very sensitive to this ratio, and even the slightest change influences ROY polymorphism. To illustrate this, one batch shown here with a solvent volume percentage of 47.00 Vol% contained yellow needles (YNs) after mixing with a magnetic stirrer, while two batches with 46.00 Vol% and 45.00 Vol% solvent presented no YNs (Figure 7b). These minimal changes with different results prompted the implementation of repetitions between 3 and 20 for the following results.

To better control mixing, a setup was created in which flows were controlled so that different parameters could be tested. Within this setup, we observed that despite the same final position within the phase diagram of ROY, the supersaturation protocol to reach this position can alter ROY polymorphism. First, the initial concentrations of the starting solution and the antisolvent solution before mixing alter the final polymorphs observed after mixing. Indeed, mixing an antisolvent mixture of 100.00 Vol% antisolvent with a starting solution until a final volume percentage of 57.50 Vol% was reached resulted in orange plates (OPs) and orange-red plates (ORPs). Contrarily, mixing a solution of 70.00 Vol% antisolvent with a starting solution until the same volume percentage of 57.50 Vol% was reached resulted in yellow prisms (Ys) (Figure 7d,e).

Secondly, different mixing methods also alter the polymorph behavior independent of the final concentration and the solvent-to-antisolvent ratio (1.0 mg/mL ROY and 42.50 Vol% solvent, respectively). Mixing by a magnetic stirrer resulted in Ys, while mixing by rotation resulted in orange needles (ONs), yellow needles (YNs), and Ys (Figure 7e,f). Similarly, mixing by rotation until an end concentration of 5.0 mg/mL ROY was reached led to a different mix of polymorphs compared with samples mixed by a magnetic stirrer (Figure 7b). The effect of the mixing rate was also analyzed: the same set of polymorphs was consistently found, indicating that the mixing rate has no influence on polymorphism (Figures S1 and S2). In conclusion, the supersaturation protocol clearly affects ROY polymorphism. These observations led to the realization that a more controlled mixing method is required and thus motivated our implementation of several mixing methods in a microfluidic setup.

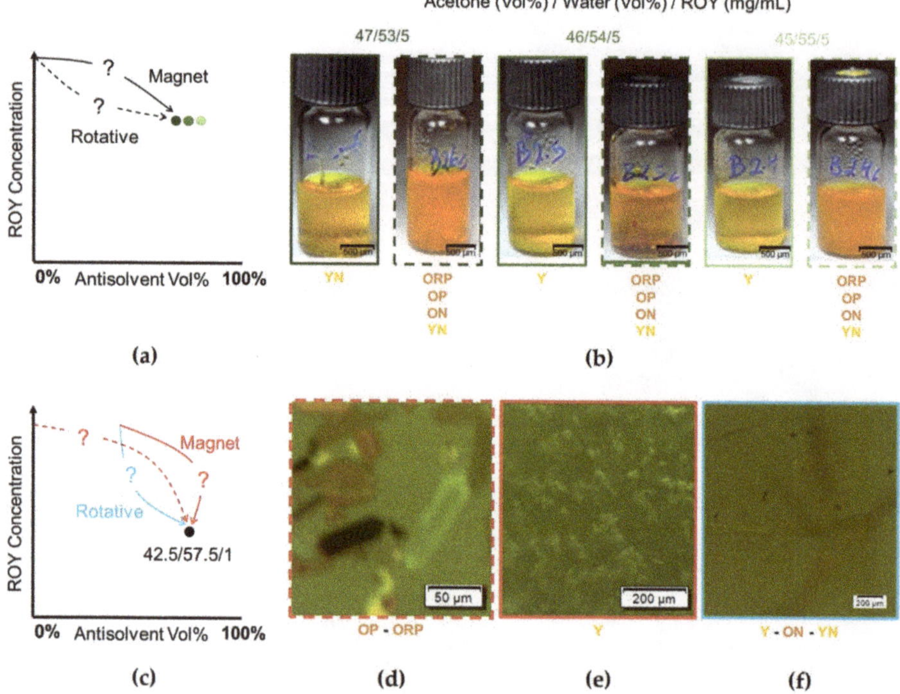

Figure 7. Supersaturation protocol influences ROY Polymorphism. (**a**) Schematic overview with possible trajectories of bulk mixing experiments performed within phase diagram; the supersaturation protocols represented here are mixing by rotation versus mixing by a magnetic stirrer. (**b**) Results after mixing with a magnetic stirrer (full line) or rotation (dotted line) with shown end concentrations and solvent to antisolvent ratios. (**c**) Schematic overview. Supersaturation protocols represented here are mixing by rotation versus mixing by a magnetic stirrer (blue versus red) and varying concentrations of the infused solutions (dotted versus full line). (**d–f**) Microscopic images of samples with the following concentrations and supersaturation protocol. (**d**) Mixed with a magnetic stirrer. Initial antisolvent solution: 100.00 Vol%. (**e**) Mixed with a magnetic stirrer. Initial antisolvent solution: 70.00 Vol%. (**f**) Mixed by rotation. Initial antisolvent solution: 100.00 Vol%. Unable to focus on all polymorphs simultaneously, the picture depicted here was carefully chosen with Ys in focus and ONs and YNs observed in the background.

3.2. *Confinement Has Minimal Effect on ROY Polymorphism*

Continuing in a microfluidic environment raises the question of whether confinement is also to be considered as a parameter that influences ROY polymorphism. To analyze this effect, samples were administered to capillaries with a diameter of 100 μm, 200 μm, or 400 μm before crystallization. The dimensions of the tested capillaries compare to the dimensions of the microfluidic chips. For all experimental conditions, the same polymorphs were found in all capillaries independently of the diameter. To mimic a more drastic change in volume, the comparison of the crystallization of identical samples between bulk and confinement was made. No major differences could be seen between bulk and confinement for the different samples tested, except for the following condition: 1.0 mg/mL ROY in 42.50 Vol% of solvent (Figure 8). In this sample, orange needles (ONs) with yellow needles (YNs) were observed when mixed in bulk, while orange plates (OPs) were observed when administered to a capillary. In order to identify the cause for these contrasting results, we set up a condition in bulk with the same solvent volume percentage and a lower ROY concentration of 0.8 mg/mL. In these conditions, OPs were formed, as they were in the capillary with 1.0 mg/mL. This suggests that the effect of confinement on ROY polymorphism is minimal and can be explained by transient local depletion of ROY molecules that can nucleate within a confined space.

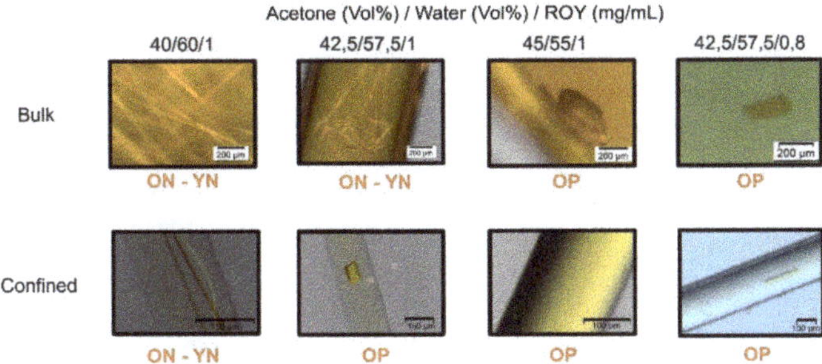

Figure 8. Confinement has minimal effect on ROY polymorphism. Obtained microscopic images of samples with concentrations and solvent to antisolvent ratios mentioned above the respective images. The upper part represents experiments performed in bulk. The bottom parts represent experiments in capillaries with an i.d. of 200 µm.

3.3. In-Chip Mixing

Two parallel flows of ROY and antisolvent solutions were infused simultaneously in the microchannel, and mixing was performed by inducing acoustic streaming. Acoustic streaming is known to be very sensitive to the experimental parameters. Indeed, acoustically soft water inside the channel surrounded by the acoustically hard silicon/glass chip forms an acoustic cavity. This implies that acoustic resonance occurs for certain specific frequencies. By tuning the applied frequency to one of the resonance frequencies so as to obtain a stationary wave along the width of the channel, the acoustic energy density inside the cavity is several orders of magnitude larger than it is at other frequencies [11], and the acoustic forces become strong enough to obtain efficient mixing [12]. The frequency of the applied signal was carefully adjusted to a resonance frequency with a precision of up to 0.001 MHz. Experimentally, this step was performed by infusing colored liquids and by monitoring the efficiency of the acoustic mixing as a function of the frequency. When an optimal frequency was reached, colored vortices were easily observed. Under these conditions, viscous attenuation of the acoustic wave in the liquid boundary layer resulted in four vortices in the direction of the acoustic propagation [11] (Figure 9). In addition, it is well known from the literature that the velocity of the liquid in the vortices increases approximately quadratically with the amplitude of the applied voltage [13]. Different mixing rates were obtained by varying the amplitude of the voltage applied to the piezoelectric element; voltage values of 0.12 V, 0.2 V, 0.5 V, and 1 V were used.

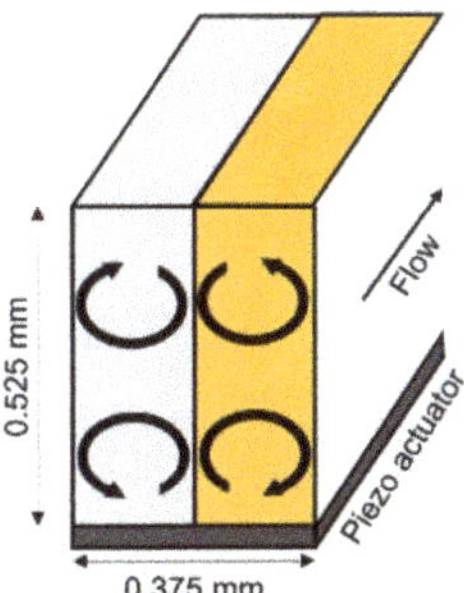

Figure 9. The frequency of the piezo element was tuned to the depth of the microfluidic channel to obtain stable vortices.

By applying sufficient acoustic power, complete mixing in the microfluidic channel occurred very rapidly. The amount of energy transferred was high, and this system caused the sample to heat up. In the absence of regulation, an increase of about 30 °C was observed within one minute. To counteract this, a PID temperature controller was used to stabilize the sample at room temperature. A typical stability of the measured temperature was ±0.1 °C (RMS value).

The two extreme cases tested are compared in Figure 10: in the first case, no voltage was applied, and the mixing occurred spontaneously by molecular diffusion. At time zero, when the flows were stopped, a clear interface between the liquids was observed. Mixing occurred gradually and was monitored qualitatively by the color variation. After about 60 s, no more color change was observed. In the second case, a 1 V_{pp} voltage was applied to the piezo element. Vortices instantaneously appeared, and complete mixing was obtained within about 2 s.

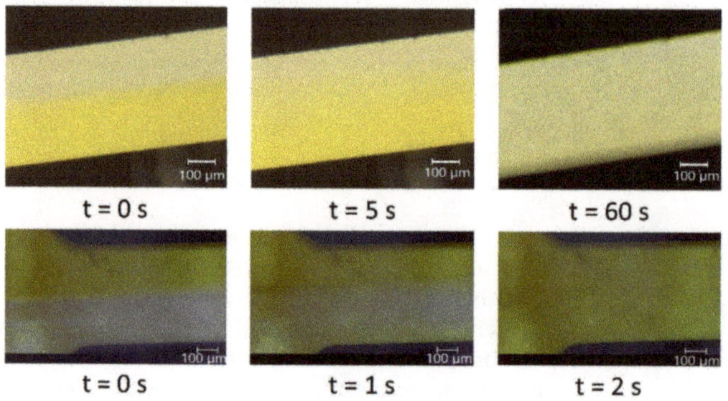

Figure 10. Acoustic mixing efficiency. **Top**: in the absence of acoustic mixing, mixing occurs through diffusion and lasts about 60 s. **Bottom**: when a 1 V_{pp} voltage is applied to the piezo electric element, vortices appear in the microchannel, and complete mixing is achieved within about 2 s.

3.4. *Supersaturation Protocol Used within Microfluidic Channel Influences ROY Polymorphism*

A large number of experiments was performed in order to investigate the polymorphism in a microfluidic environment. Two main parameters were tested: on the one hand, different solvent and antisolvent concentration gradients were tested by varying the concentration of the infused solutions. For the first series of samples, depicted with dotted lines in Figure 11, pure antisolvent and ROY–acetone solutions were infused in the channel. For the other samples, depicted with solid lines in Figure 11, smaller solvent and antisolvent concentration gradients were created by infusing solutions with 70.00 and 55.00 Vol% antisolvent. Respective flowrates were adjusted so that after mixing, the final concentration in the microchannel corresponded to 42.50 Vol% acetone, 57.50 Vol% water, and 1.0 mg/mL ROY. These concentrations were identical to the concentrations of samples d, e, and f tested in bulk conditions in Figure 7.

Both the initial concentrations of the solutions and the mixing methods were found to play an important role in polymorphism. Indeed, large initial concentration gradients (Figure 11a) and diffusive mixing consistently resulted in yellow and orange needles (YNs and ONs), as shown in Figure 11b. The needles appeared in the central region of the channel, at the interface between the two liquids, and quickly developed in the entire volume.

However, with the same solutions, crystallization appeared to be very different when acoustic mixing was applied and mixtures of orange plates (OPs) and yellow prisms (Ys) were obtained for all tested samples, as shown in Figure 11c.

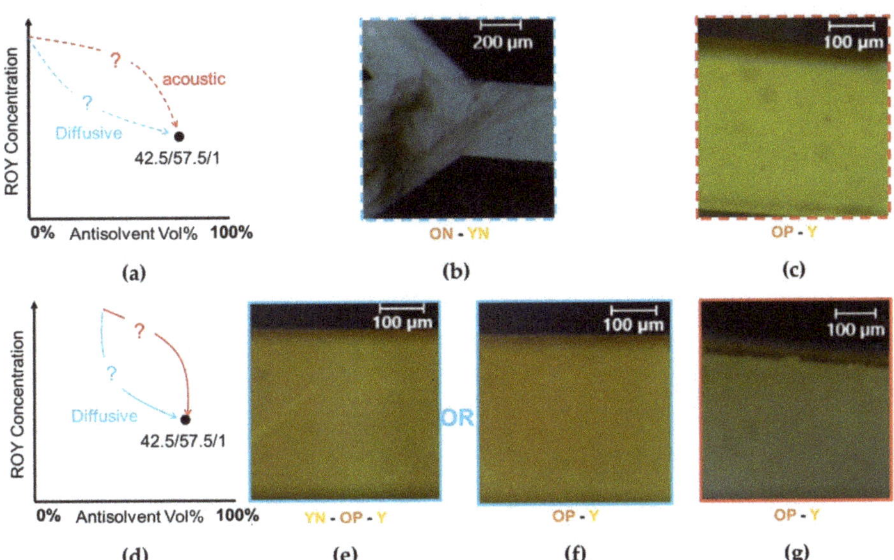

Figure 11. The supersaturation protocol influences ROY polymorphism in a microfluidic environment. Unable to focus on all polymorphs simultaneously, the pictures depicted here were carefully chosen with Ys in focus and the other polymorphs observed in the background. (**a**,**d**) Two supersaturation protocols were tested by varying the concentrations of the infused solutions. Influence of mixing rate was investigated by comparing diffusive mixing and acoustic mixing. (**b**) Larger initial concentration gradients and diffusive mixing consistently resulted in orange needles (ONs) and yellow needles (YNs). (**c**) Larger initial concentration gradients and acoustic mixing consistently resulted in orange plates (OPs) and yellow prisms (Ys). (**e**,**f**) Smaller concentration gradients and diffusive mixing produced different results: in 7 samples out of 10, orange plates (OPs) and yellow prims (Ys) were obtained. In 3 samples, yellow needles (YNs) were also observed. (**g**) Smaller initial concentration gradients and acoustic mixing consistently resulted in orange plates (OPs) and yellow prisms (Ys).

With smaller concentration gradients (Figure 11d), diffusive mixing resulted in different polymorphs. For most samples (7 out of 10), a mixture of OPs and Ys was obtained (Figure 11f), while for three samples, large YNs were also observed (Figure 11e). Orange plates were mostly observed along the walls of the channel, but crystals appeared everywhere. Finally, for those solutions and in all tested samples, acoustic mixing resulted in OPs and Ys, as shown in Figure 11g.

In these samples, crystallization began as soon as mixing started along the walls of the microchannels, and crystals were observed in the entire volume in a few seconds. In most experimental runs, OPs grew faster, and Ys were detected 30 s to one minute after the mixing. Identical polymorphs were obtained for all tested intensities of acoustic mixing (Figure S3).

4. Discussion

From the overall experimental dataset obtained in this study, we made some general key observations about ROY polymorphism. ROY polymorphism is very sensitive to the solvent-to-antisolvent ratio, as shown in the results obtained in Figure 7. Moreover, it appears that mixing is a key parameter. A homogeneous sample after mixing by rotation is reached after several seconds. On the timescale of seconds, different local concentrations are possible in the vial during the entire infusion process, and strong local concentration gradients exist in the mixture. This allows the system to sample several regions in the free energy diagram and nuclei of various polymorphs to form and reach critical sizes. This results in very complex final mixtures of many polymorphs, as shown in the dotted lined samples in Figure 7b. This statement is supported by the results displayed in Figure 7d,e, in which different start concentrations were used to mix to the final end concentration. Indeed, lowering the initial concentration gradients results in smaller local concentration gradients, resulting in a smaller variety of polymorphs in the end. Contrarily, mixing with a magnetic stirrer allows faster mixing and keeps the concentration homogeneous in the entire vial. This mixing method results in less drastic local concentration gradients, prohibiting the system from reaching certain regions in the free energy diagram. This gives rise to a smaller variety of polymorphs compared with the samples mixed by rotation.

Similar observations emerged when analyzing the results of microfluidic experiments. The results in Figure 8 show that the influence of the confinement of the sample is low. However, for identical final conditions, two mixing methods within the microfluidic channels led to other polymorphs, as shown in Figure 11. Diffusive mixing is slower, and the concentration gradients in the liquid result in less reproducible experiments allowing the formation of different polymorphs, while acoustic mixing always leads to orange plates and yellow prisms. Acoustic mixing is very efficient, and the system reaches its final mean concentration in the entire microfluidic channel within 1 to 2 s. These results indicate that in the final concentration condition, that is, 47.50 Vol% acetone, 52.50 Vol% water, and 1.0 mg/mL ROY; orange plates and yellow prisms are the most stable polymorphs.

It also appears that needles can be obtained with rotative mixing in bulk or with diffusive mixing in a microchannel, but over the very large number of experiments that we performed, needles were never obtained with magnetic stirring in vials nor with acoustic mixing in microchannels. These observations suggest that needle nucleation occurs on longer timescales. On the basis of our experiments, however, it is not possible to draw a solid conclusion, and the formation of needles could also be hindered by excessive shear and flow resulting from the magnetic stirring and acoustic mixing. Moreover, a separate study on nucleation may bring interesting insights into the driver of these differences observed.

Throughout all our experiments, orange-red plates were observed in very few samples. This polymorph is not stable in the concentration domain that we tested. Finally, it appears that the microfluidic environment and the use of fast mixing methods represent an interesting opportunity for the selective production of certain polymorphs. Under the conditions that we tested, the preparation of the samples in microfluidic reactors and using acoustic mixing consistently produced orange plates and yellow prisms.

5. Conclusions

This work reports on ROY crystallization in various conditions. Not only do the final end concentration of the solvent, antisolvent, and ROY affect polymorphism, but the selected protocol used to reach supersaturation also affects the polymorphism behavior. Here, we reported on two different sets of protocols resulting in different polymorphs: first by changing the initial mixing solutions, and second, by altering the mixing method. In bulk, mixing was performed either by rotation or a magnetic stirrer. In a microfluidic setup, we created controlled mixing environments and tested the two sets of protocols: changing the initial concentrations and the mixing method. In the microfluidic channel, this was achieved by either diffusive mixing or mixing acoustic streaming. To analyze mixing in a microfluidic setup, the influence of confinement on polymorphism was considered. The only effect of confinement that we discovered was the transient local depletion of ROY molecules.

Both in bulk and in a microfluidic environment, changing the initial stock solutions resulted in different polymorphs, as did the applied mixing method. We believe that mixing by rotation in bulk and mixing by diffusion in a microfluidic channel allows the system to reach minima in the free energy diagram because of slow mixing and large local gradients, which are not reached (or do not last long enough) when fast mixing is performed by a magnetic stirrer in bulk or acoustic streaming in a microfluidic channel. This results in more variable and less stable polymorphs in the conditions tested when slow mixing occurs. On the basis of these findings, we conclude that orange plates and yellow prisms are the most stable polymorphs in the system that we tested. Indeed, needles only appear under gentle mixing conditions, such as diffusion and mixing by rotation, while orange-red plates were sporadically obtained.

In the microfluidic setup, in which mixing can be accurately controlled, orange plates and yellow prisms can be efficiently and reproducibly selected. Overall, these findings show that the start conditions, supersaturation protocol, and flow are critical parameters in ROY polymorphism that have to be considered in the design of systems for polymorph selectivity.

Supplementary Materials: The following supporting information can be downloaded at: https://www.mdpi.com/article/10.3390/cryst12050577/s1, Figure S1: mixing rate in bulk experiments with rotative mixing does not influence polymorphism; Figure S2: mixing rate in bulk experiments with magnetic stirrer does not influence polymorphism; Figure S3: intensity of acoustic mixing does not influence polymorphism in microfluidics.

Author Contributions: Conceptualization, D.M., W.D.M., Q.G. and M.V.N.; methodology, D.M., Q.G., M.H. and M.V.N.; experiment design, Q.G., M.V.N., M.H. and P.G.; resources, M.H.; writing—original draft preparation, M.V.N. and Q.G.; writing—review and editing, M.V.N., Q.G., D.M. and J.F.L.; supervision, D.M. and W.D.M.; project administration, D.M.; funding acquisition, D.M. All authors have read and agreed to the published version of the manuscript.

Funding: This research was funded by the European Space Agency under Prodex Contract No. ESA AO-2004-070 and by a Strategic Research Program on Microfluidics (SRP51) at Vrije Universiteit Brussel.

Institutional Review Board Statement: Not applicable.

Informed Consent Statement: Not applicable.

Data Availability Statement: Not applicable.

Conflicts of Interest: The authors declare no conflict of interest. The funders had no role in the design of the study; in the collection, analyses, or interpretation of data; in the writing of the manuscript; or in the decision to publish the results.

References

1. Morrow, S. One hundred and fifty years of isomorphism. *J. Chem. Educ.* **1969**, *46*, 580. [CrossRef]
2. Cruz-Cabeza, A.; Bernstein, J. Conformational polymorphism. *Chem. Rev.* **2014**, *114*, 2170–2191. [CrossRef] [PubMed]
3. Lee, E. A practical guide to pharmaceutical polymorph screening & selection. *Asian J. Pharm. Sci.* **2014**, *9*, 163–175. [CrossRef]
4. Thakuria, R.; Thakur, R.S. Crystal Polymorphism in Pharmaceutical Science. In *Comprehensive Supramolecular Chemistry II*, 2nd ed.; Atwood, J.L., Ed.; Elsevier: Edinburgh, UK, 2017; Volume 2, pp. 283–309.
5. Raza, K. Polymorphism: The Phenomenon Affecting the Performance of Drugs. *SOJ Pharm. Pharm. Sci.* **2014**, *1*, 10. [CrossRef]
6. Mitchell, C.A.; Yu, L.; Ward, M.D. Selective nucleation and discovery of organic polymorphs through epitaxy with single crystal substrates. *J. Am. Chem. Soc.* **2001**, *123*, 10830–10839. [CrossRef] [PubMed]
7. Gushurst, K.S.; Nyman, J.; Boerrigter, S.X.M. The PO13 crystal structure of ROY. *Cryst. Eng. Comm.* **2019**, *21*, 1363–1368. [CrossRef]
8. Yu, L. Polymorphism in molecular solids: An extraordinary system of red, orange, and yellow crystals. *Acc. Chem. Res.* **2010**, *43*, 1257–1266. [CrossRef] [PubMed]
9. Nagaki, W.; Doki, N.; Yokota, M.; Yamashita, K.; Kojima, T.; Tanaka, T. Control of Crystal Size and Morphology of Calcium Carbonate Crystal Polymorphism. *J. Mater. Sci. Chem. Eng.* **2021**, *9*, 38–45. [CrossRef]
10. Ziemecka, I.; Gokalp, S.; Stroobants, S.; Brau, F.; Maes, D.; de Wit, A. Polymorph selection of ROY by flow-driven crystallization. *Crystals* **2019**, *9*, 351. [CrossRef]
11. Barnkob, R.; Augustsson, P.; Laurell, T.; Bruus, H. Measuring the local pressure amplitude in microchannel acoustophoresis. *Lab Chip* **2010**, *10*, 563–570. [CrossRef] [PubMed]
12. Gelin, P.; Maes, D.; de Malsche, W. Reducing Taylor-Aris dispersion by exploiting lateral convection associated with acoustic streaming. *Chem. Eng. J.* **2021**, *417*, 128031. [CrossRef]
13. Muller, P.B.; Barnkob, R.; Jensen, M.J.H.; Bruus, H. A numerical study of microparticle acoustophoresis driven by acoustic radiation forces and streaming-induced drag forces. *Lab Chip* **2012**, *12*, 4617–4627. [CrossRef] [PubMed]

Article

Crystal Breakage Due to Combined Normal and Shear Loading

Benjamin Radel *, Marco Gleiß and Hermann Nirschl

Institute of Mechanical Process Engineering and Mechanics, Karlsruhe Institute of Technology, 8 Strasse am Forum, 76137 Karlsruhe, Germany; marco.gleiss@kit.edu (M.G.); hermann.nirschl@kit.edu (H.N.)
* Correspondence: benjamin.radel@kit.edu

Abstract: Combined normal and shear stress on particles occurs in many devices for solid–liquid separation. Protein crystals are much more fragile compared to conventional crystals because of their high water content. Therefore, unwanted crystal breakage is to be expected in the processing of such materials. The influence of pressure and shearing has been investigated individually in the past. To analyze the influence of combined shear and normal stress on protein crystals, a modified shear cell for a ring shear tester is used. This device allows one to accurately vary the normal and shear stress on moist crystals in a saturated particle bed. Analyzing the protein crystals in a moist state is important because the mechanical properties change significantly after drying. The results show a big influence of the applied normal stress on crystal breakage while shearing. Higher normal loading leads to a much bigger comminution. The shear velocity, however, has a comparatively negligible influence.

Keywords: protein crystals; breakage; shear stress

Citation: Radel, B.; Gleiß, M.; Nirschl, H. Crystal Breakage Due to Combined Normal and Shear Loading. *Crystals* **2022**, *12*, 644. https://doi.org/10.3390/cryst12050644

Academic Editor: Abel Moreno

Received: 31 March 2022
Accepted: 27 April 2022
Published: 30 April 2022

1. Introduction

In recent years, alternative methods to purify and formulate proteins have been investigated. One example is preparative protein crystallization [1], which provides the opportunity to replace a costly chromatography step with selective protein crystallization. Crystalline proteins offer several advantages. By influencing the crystal shape and size, product characteristics like handling, shelf life, and drug release properties can be adjusted [2,3]. One well-known example is the use of crystalline insulin to achieve a depot effect and constant bioavailability [4]. Conventionally, protein crystallization on a larger scale uses the displacement method. Added anti-solvents (typically salts) reduce the target protein's solubility and, thus, create the required supersaturation. Inherently, this method has some major disadvantages. First, the anti-solvent is typically added as a solution, which dilutes and, thus reduces the final crystal concentration in the suspension. Second, the final solution has high ion strength and, finally, high local supersaturation at the anti-solvent inlet leads to a broad crystal size distribution, lower reproducibility, and increased danger of amorphous precipitation [5]. Evaporative crystallization tackles these disadvantages, but is not suitable for proteins because of the required high temperatures. Groß and Kind [5] introduced low temperature water evaporation crystallization for proteins and demonstrated the applicability for the model system lysozyme from hen-egg white. Small scale filtration experiments of such crystallizate by Radel et al. [6,7] allowed to determine the process functions of this particle system. Using these process functions, the low temperature water evaporation crystallization method was adapted and implemented by Dobler et al. [8] for an integrated, quasi-continuous apparatus concept. Barros Groß and Kind [9] further investigated how seeding affects the crystal size distribution and demonstrated high reproducibility and control over crystal sizes and the width of the size distribution.

One major difference between protein and conventional crystals is the high water content in the protein crystal, which can account for up to 80% of the crystal mass. This makes such crystals soft and sensitive to mechanical stress [10]. Cornehl et al. [11] investigated crystal breakage of aggregated and needle-shaped lysozyme crystals under compressive stress. In process engineering, compressive stress occurs, for example, in press or cake filtration. Other apparatuses for filtration, such as the cross-flow filter (BoCross Dynamic, BOKELA GmbH, Karlsruhe, Germany) are characterized by high shear stress due to integrated agitators. In these cases, Cornehl et al. [12] also observed crystal comminution. In crystallization itself, comminution is partly a desirable effect. For example, collision of the crystals with each other as well as with internals, walls, and stirrers creates new crystallization nuclei that subsequently grow into larger crystals. For the use of crystallization as a formulation step for downstream processing of proteins [3,13], comminution after crystallization is usually not desired, since a change in particle size distribution (PSD) also changes the product properties.

Sediments can be compacted by both normal and shear stress. If both types of stress are combined, the achievable compaction is more pronounced, as used, for example, by Illies et al. [14] for the dewatering of filter cakes. Höfgen et al. [15] used an apparatus with high pressure dewatering rolls for filtration and dewatering. With this system, the shear and normal forces are individually adjustable and adaptable to the application at hand. Hammerich et al. [16] modified a shear cell for the Schulze ring shear tester to allow the measurement of fluid saturated particulate networks. This provides the opportunity to study the rheology and flow behavior of saturated sediments. The higher achievable compaction with combined shear and compression indicates bigger mechanical stress. Hence, a more pronounced comminution for protein crystals is to be expected in this case. The setup with the modified shear cell can, therefore, be used to observe comminution at a defined normal stress and shear velocity.

2. Theory

Particle breakage occurs when the material strength is exceeded. In addition to a one-time high mechanical load, fracture due to several lower load cycles is also conceivable. In this case, the loading history of the particle must be taken into account. For comminution processes such as grinding, the occurrence of particle or aggregate breakage is the basic requirement. In most solid–liquid separation applications, particle breakage is undesirable and usually leads to a deterioration of the process result. Basically, a distinction must be made between the stress in a particulate network and on the individual particle. In the following, the special focus is on compressive and shear stress.

In a particulate network under compressive stress, the imposed mechanical load is degradable by rearrangement processes, deformation, and fracture of the particles. For an axial, one-dimensional load, such as that applied by a piston, the effects mentioned above result in the highest load being applied in the immediate vicinity of the piston. The load decreases with increasing distance from the piston in the sediment. If the sediment is compressible, the highest compaction is to be expected in the vicinity of the piston [17]. In the upper particle layers, absorption of the input mechanical energy takes place due to elastic and plastic deformation and particle breakage, such that the mechanical energy transmitted via contact points decreases. Thus, the lowest stresses are present at the greatest distance from the piston center and at the bottom [18].

Shear stresses occur in process engineering, for example, on agitators, valves, and in pumps. Fluid-induced mechanical stress occurs in a suspension due to the existing flow. In addition to the fluid-induced stress, collisions between particles and particles with agitators or walls take place. These collisions lead to particle abrasion and, at high loads, to particle breakage. In crystallization, such abrasion is partly desirable, as it creates new crystallization nuclei for secondary nucleation.

Shear loading is also possible for saturated sediments. This applies to the transport of flowable sediments in equipment, such as the tubular centrifuge or decanter centrifuge, as well as to the targeted use of combined normal and shear loading for mechanical dewatering. When normal and shear forces are superimposed, significantly higher densities and, thus, lower residual moisture content can be achieved [14,15]. In such scenarios, the particles rub against each other and, in the case of mechanically labile particle systems, abrasion or fracture occurs.

Comminution

The PSD, particle shape, and sediment structure have a big influence on the energy absorption and stress distribution in the particle bed. Furthermore, the overall stress is affected by the normal and shear stress as well as the shear velocity. As stress increases, comminution occurs in addition to compaction. Comminution leads to new fracture surfaces and rearrangement processes. The smaller the particles, the higher the energies required to cause comminution.

Population balances can be used to model the comminution of different particle size classes. The probability of breakage, P, during impact loading is given for many materials by the exponential function

$$P = 1 - \exp(-fxk(E_M - E_{M,\min})) \tag{1}$$

with f as the material parameter, x as the particle diameter, k as the number of load cycles, E_M as mass-related stress energy, and $E_{M,\min}$ as the threshold value of the mass-related stress energy [18]. The breakage fraction $\bar{P}$ is identical to the breakage probability for a single grain. In a collective, the breakage fraction can be described by normalization with the master curve

$$\frac{\bar{P}}{\bar{P}_\infty} = 1 - \exp\left(-\left(\frac{E_M}{E_{M,c}}\right)^\beta\right). \tag{2}$$

The quantity $\bar{P}_\infty$ is the upper limit of $\bar{P}$ and β is a curve parameter. $E_{M,c}$ is a characteristic value of E_M for which different approaches, taking into account the particle size, exist [19].

A fracture force can be determined on a single grain via compression or indentation tests. Nanoindentation is a suitable method for small particles that are to be examined moist, as in the case of protein crystals. Depending on the particle shape, different loads are conceivable. For an elongated particle, for example, a three-point bending test is possible. To cause particle fracture in a particulate network, a higher force than the fracture force of the single particle is necessary. This is due to the absorption of mechanical stress by rearrangement and deformation processes. Particle breakage is divided into the following phases:

1. cracking;
2. crack initiation;
3. crack propagation.

Cracking starts as soon as the material strength is exceeded. Therefore, cracks preferentially form at locations that are already under stress or where defects are present. Defects can be, for example, lattice defects in a crystal. When the energy is low, a dormant crack is formed in this way. If no further stress takes place, the crack does not propagate and no fracture occurs. Only at a sufficiently high energy does crack initiation and subsequent crack propagation occur. The energy required for crack propagation at the cracking front must be continuously replenished.

Therefore, two conditions apply to particle breakage. On the one hand, there is the force condition, i.e., overcoming the binding forces for the formation of a crack, and on the other hand, there is the energy condition. The latter states that energy consumed at the crack front must be continuously supplied [20].

3. Materials and Methods

3.1. Crystallization

Isometric lysozyme crystals are produced with displacement crystallization. Two stock solutions are prepared. Solution one is an 25 mmol L^{-1} acetic acid buffer at pH 4.5. Solution two has the same composition but also contains 80 g L^{-1} NaCl. After dissolving 100 g L^{-1} lysozyme (Granulated lysozyme, OVOBEST Eiprodukte GmbH & Co. KG, Neuenkirchen-Vörden, Germany) from hen egg-white in 125 mL of stock solution one, 125 mL of stock solution two is added at a rate of 1 mL min^{-1} with a membrane pump. Afterwards, the crystals grow in the aging phase until the supersaturation is reduced to zero. During the whole crystallization process, which takes about 16 h, the solution is stirred with a blade stirrer at 350 min^{-1}.

3.2. Modified Shear Cell and Ring Shear Tester

For the combined normal and shear loading, sediments of isometric lysozyme crystals are used as a model system for protein crystals. The mechanical properties of dried and moist crystals differ greatly. Therefore, for a realistic assessment of the material behavior, it is important to measure the crystals as close as possible to the actual process conditions. To load the crystals in a saturated sediment, the ring shear tester (RST-01.pc, Dr. Dietmar Schulze) is used. This device is common in the field of bulk mechanics to characterize the flow properties of dry powders. The bulk material is placed in a shear cell. The shear cell lid with attached drivers is placed on the bulk material and hooked onto a weight. The normal stress of the system is determined by the applied weight. For shearing, the shear cell rotates, but the shear cell lid is connected and fixed to bending beams via tie rods. The shear stress can, therefore, be measured at the bending beams.

Modifications to the shear cell are necessary for the measurement of moist sediments. A detailed explanation of the modified shear cell can be found in Hammerich et al. [16]. Therefore, the modifications are only roughly outlined in the following. When moist sediments are loaded, compaction results in the displacement of fluid, which must escape from the system. In contrast, when the sediment is stretched back, it must be ensured that no desaturation occurs.

The structural implementation of the ring-shaped modified shear cell is shown as a sectional view in Figure 1. The shear cell lid consists of two parts, which are sealed against each other by Teflon rings. To allow liquid to escape, but at the same time retain the particles, filter media with a support structure are installed on the top and bottom of the shear cell, shown in green in Figure 1. The drivers and the bottom plate of the shear cell are visualized in gray. On the bottom of the shear cell, below the filter medium, a drainage channel is located, which allows the displaced fluid to escape. The L-shaped profile ensures that the drainage channel is always filled with liquid. Thus, the sediment does not desaturate in the event of back expansion. For the same purpose, there is a riser tube for the displaced liquid on the shear cell lid.

Due to the additional Teflon seals and the slightly different design, the measured values for normal and shear stress cannot be determined directly. A correction for the friction contributions of the seals is required. For this purpose, Hammerich et al. [16] developed a variant of the shear cell with strain gauges for load measurement and determined correction functions for the normal and shear stress. These functions can be found in [16]. The correction functions depend on the seal set used and, therefore, must be recalibrated when replacing the seals.

An experiment with the ring shear tester and the modified shear cell is divided into the following phases:

1. sample preparation;
2. shear cell assembly;
3. shear cell preparation;
4. shear test.

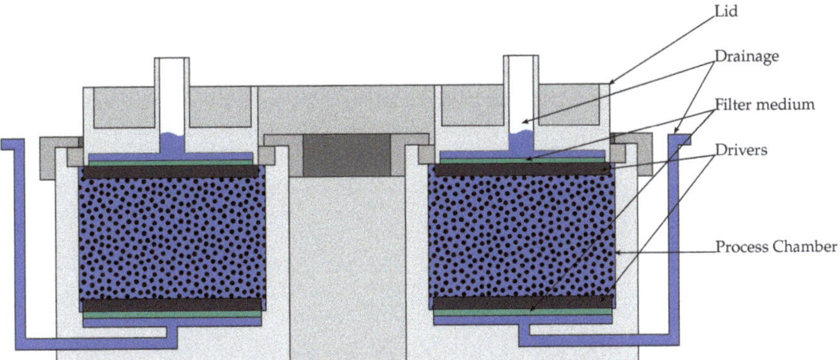

Figure 1. Sectional view of the modified shear cell; adapted from [16].

Sample preparation consists of preparing isometric lysozyme crystals via displacement crystallization. To provide a sufficient amount of crystallizate, and to compensate for minor variations between the different batches, three individual crystallization batches have to be prepared and mixed. The crystals settle overnight. The following day, the supernatant can be decanted. This increases the concentration of crystallizate when the sediment is resuspended and the higher solids volume fraction prevents segregation due to zone sedimentation.

When assembling the shear cell, the filter medium (Trakedge 0.2 µm, Sabeu GmbH & Co. KG, Northeim, Germany) the support structures, and the base plate have to be installed in the bottom part of the shear cell. The shear cell bottom is placed into a customized centrifuge insert for the beaker centrifuge (ZK 630, Berthold Hermle AG, Gosheim, Germany). This insert allows the sediment to be built up by centrifugation directly in the shear cell. Suspension is added to the beaker, so that the processing chamber of the shear cell bottom is slightly overfilled. The sample is centrifuged at a speed of $1500\,\mathrm{min}^{-1}$ for 10 min. This corresponds to a maximum relative centrifugal acceleration of $C = 500$. Due to the comparably low centrifugal acceleration, the normal loading of the particulate network is low and a change in PSD due to centrifugation can be avoided. After removal from the centrifuge, the sediment has been built up in the shear cell processing chamber. The supernatant is discarded carefully. The L-shaped drainage channels are filled with supernatant. To ensure a defined sediment level for the different tests, excess sediment is carefully removed using a scraper with a fixed length.

The same filter membrane, support structures, and drivers must also be mounted on the shear cell lid. The shear cell lid consists of two parts and, therefore, requires sealing with Teflon O-rings. The assembled lid is centered and placed on the shear cell with the sediment. By softly pressing on the lid, it is fixed in the correct orientation.

The shear cell preparation takes place next. The assembled shear cell is inserted into the ring shear tester device, the tie rods are installed, and the counterweight is attached to the lid. To finally position the shear cell lid in a defined way, a normal load of 40,180 Pa is applied. This causes the lid to slide into position and displace some fluid from the sediment. The riser tubes are then filled with supernatant. Afterwards, the normal load is removed, and the shear cell is now ready for the actual shear test.

The shear test consists of applying the defined normal stress and then shearing the specimen by rotating the bottom part of the shear cell. The adjustable parameters are the shear velocity, the normal stress, and the shear duration. For the characterization of powders, the default shear velocity is $1.5\,\mathrm{mm\,min}^{-1}$. The velocity is varied in the range of 0.48–$4.5\,\mathrm{mm\,min}^{-1}$ for the combined shear and normal loading of the moist protein crystals. In the experiments, the shearing time is varied in such a way that the shearing path remains constant. A schematic plot of the shear stress versus shear duration is shown in Figure 2. After removal of the sheared sediment from the shear cell, samples for measuring the PSD with laser scattering are taken and resuspended in supernatant. Additionaly, samples are cut out of the sediment with a polyimide tube for characterization with micro computed tomography (µCT). To assess the influence of centrifugation and the compaction at shear cell preparation on the PSD, samples of the initial solution and after centrifugation and compaction without shearing in the ring shear apparatus are also analyzed.

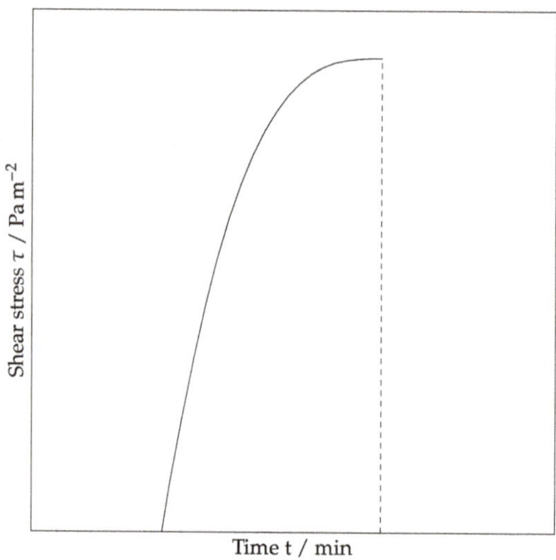

Figure 2. Schematic representation of the shear stress curve.

3.3. *Micro Computed Tomography Analytic*

The working principle and sample preparation as well as post processing are explained in detail in Dobler et al. [8]. Thus, the preparation of the samples is only briefly outlined in the following. The polyimide tubes holding the samples of the compressed and sheared sediment are deep-frozen with liquid nitrogen. Afterwards, the pore fluid is removed with lyophilization. The removal of pore fluid ensures better contrast between particles and the surrounding air. For the measurement in the µCT (Zeiss Xradia Versa 520, Carl Zeiss Microscopy GmbH, Oberkochen, Germany), the sample tube has to be glued onto a dress pin and mounted on a sample holder. At an energy of 50 W, 2201 projections (X-ray images) are taken from the sample at different rotation angles. These projections are reconstructed to a 3D 16 grayscale image stack. The resulting voxel size is about 400 nm.

4. Results and Discussion

The shear stress applied in the shear cell has a gradient and, thus, varies in the sample. Directly at the shear cell lid, between the drivers protruding into the sample, the shear stress is zero. Below the drivers, the highest load occurs, which then decreases towards the bottom of the shear cell. This also results in differences with regard to the comminution that occurs. No comminution takes place directly between the drivers of the lid.

The resulting shear zones are also visible in µCT images of the loaded sediment. Figure 3 shows an image of the top of the sediment after shearing with a normal stress of 88,396 Pa. No comminution, but only compression, takes place in the uppermost layer of the sediment (green). This is the area between the approximately 2 mm long drivers, which protrude into the sample. When disassembling the shear cell, a small amount of the sediment sticks to the shear cell lid. Thus, the height of the undamaged crystal layer is approximately 600 µm. The impression of these drivers in the sediment are also visible on the top of the µCT images in Figure 4. The particles in the area between the grooves of the top layer in Figure 4 experience no shear force. In Figure 3, the sharply delineated, red-colored layer directly below has the highest shear force. The resulting comminution of the crystals is so strong, that the crystal structure can no longer be visualized with the µCT. The particle sizes and the porosity is too low for the CT scan resolution. This is shown by the fact that practically no crystal edges can be seen.

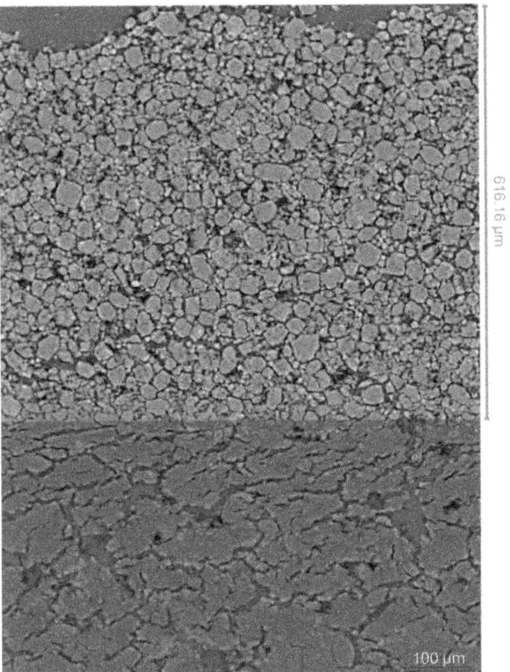

Figure 3. Shear zones in the sediment: **Top** (green), undamaged crystals in the area between the drivers; **bottom** (red), crushed and no longer resolvable crystals.

Figure 4. Computer tomography image of a bulk (gray) with tracer particles (green) in the initial unloaded (**left**) and sheared (**right**) state.

Comminution at the bottom of the shear cell and at the very top between the drivers is much less pronounced than at the top, directly below the drivers, of the saturated sediment. This fact is also confirmed by µCT scans of a model sediment in the initial unloaded and sheared condition in a miniaturized shear cell. The visualization of these scans is shown in Figure 4. The particles colored in green are a model system with lower density, which therefore have a different gray value in the tomography. Thus, the initial and final position of these particles can be marked. The grooves seen on the top of the sediment are from the drivers. To better visualize the green particles, the other gray material in the front area is shown faded. In the initial state on the left side of Figure 4, the green tracer particles are present in a vertical stripe in the sediment. After a shear path of about 45°, the distribution of the tracer particles shown on the right is obtained. A large part of the tracer particles remains unsheared and stays together in a strip. In the upper region near the lid, however, there is a clear distribution of the tracer material along the shear path. In this area, the particles experience high shear stress, which, in the case of mechanically sensitive materials, also causes comminution. The distribution of the particles also explains why not all large particles are crushed. In the lower area of the sediment and at the top between the drivers, basically no comminution takes place.

Figure 5 shows the PSD of the unloaded suspension and of two layers of the sediment after shearing. The comminution of the sheared samples compared to the unloaded initial suspension can be seen. Furthermore, it can be seen from Figure 5, that the PSD of the top layer of the bulk has a higher fines content than the lower layer and the total sediment. The sheared total curve in Figure 5 is a laser diffraction measurement of a resuspendend sample of the sediment with complete sediment height. This curve, therefore, represents all layers in the sample. The comminution is particularly pronounced for the x_{10} diameter, which is only between 10–15% of the unloaded suspension. The x_{90} diameter decreases much less and is in the range between 75–85%. This can be explained by the aforementioned unequal shear stress in the sediment. Large particles remain intact at the bottom of the bulk and at the same time an increase in very small particles is observed. Thus, the PSD broadens.

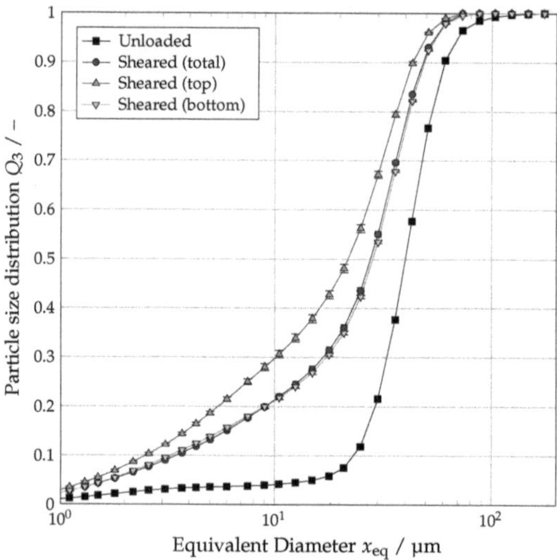

Figure 5. Particle size distributions in the different bulk layers after shearing.

Various variables influencing comminution are conceivable. In particular, the applied normal stress and the shear velocity probably have an effect on crystal breakage. These two influencing variables are, therefore, considered in more detail in this study. The covered shear path for all experiments remains a constant 15 mm in order to exclude the influence of different shear paths. For the evaluation of the changes of the PSD, the comparison between the respective unloaded suspension and a resuspended sample of the total sediment after loading is used. The PSD is obtained with laser diffraction.

4.1. Influence of Centrifugation and Compaction

It is conceivable that centrifugation and compaction for sample preparation in the ring shear tester also causes particle size reduction. To assess this influence, the PSDs of unloaded and centrifuged and compacted samples are shown in Figure 6. Compacted in this context means an applied normal load of the sediment in the ring shear tester with σ_n = 40,180 Pa, without shear loading and after centrifugation.

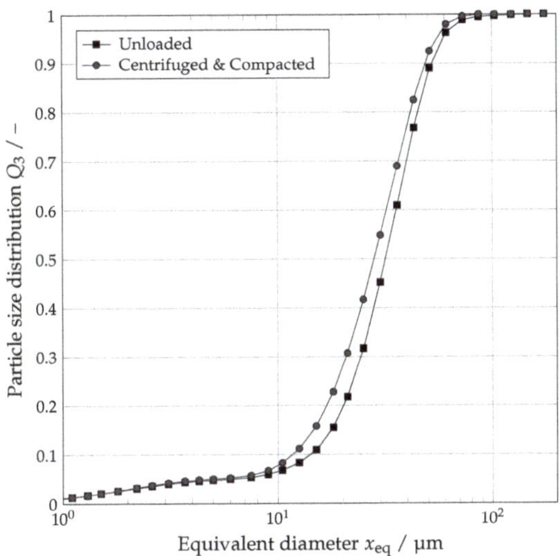

Figure 6. Comparison of particle size distributions for the unloaded suspension and centrifuged and compacted sediments.

The resulting PSDs can be seen in Figure 6. The crystals in Figure 6 are centrifuged at a speed of 1500 min^{-1} and subsequently compacted in the ring shear tester. Afterwards, the sediment has been resuspended in supernatant for the determination of the PSD with laser diffraction. They show a slight shift towards smaller particle sizes. However, the whole PSD is shifted and no increase in fines is observed. The shift is so weakly pronounced that comminution is negligible compared to the sheared samples. This slight change in PSD can also be explained by minor temperature variation during measurement and is not necessarily due to centrifugation. Hence, the preparation of the shear cell and the subsequent sampling, does not affect the PSD significantly. In summary, therefore, neither the load in the centrifuge nor the additional normal load in the shear cell causes significant particle size reduction or crystal breakage.

4.2. Influence of Normal Stress

Shear tests at a constant shear velocity of 1.5 mm min^{-1} and varying normal stresses are used to evaluate the influence of the applied normal stress on comminution. The transformed particle size density distribution q_3^* for varying normal stress is shown to visualize the comminution. Since this is a biological system that has natural fluctuations during crystallization, the initial distribution is also displayed for each normal stress.

The resulting PSDs are shown in Figure 7. Already at a low normal load of 20,572 Pa a size reduction, as well as a shift of the particle sizes, can be observed and the distribution width increases by about 40%. The x_{90} and x_{50} diameters decrease significantly at this low load to 78 and 68% compared to the unloaded particle sizes. The most significant decrease is observed for the x_{10} diameter. This diameter is reduced to 30% after loading.

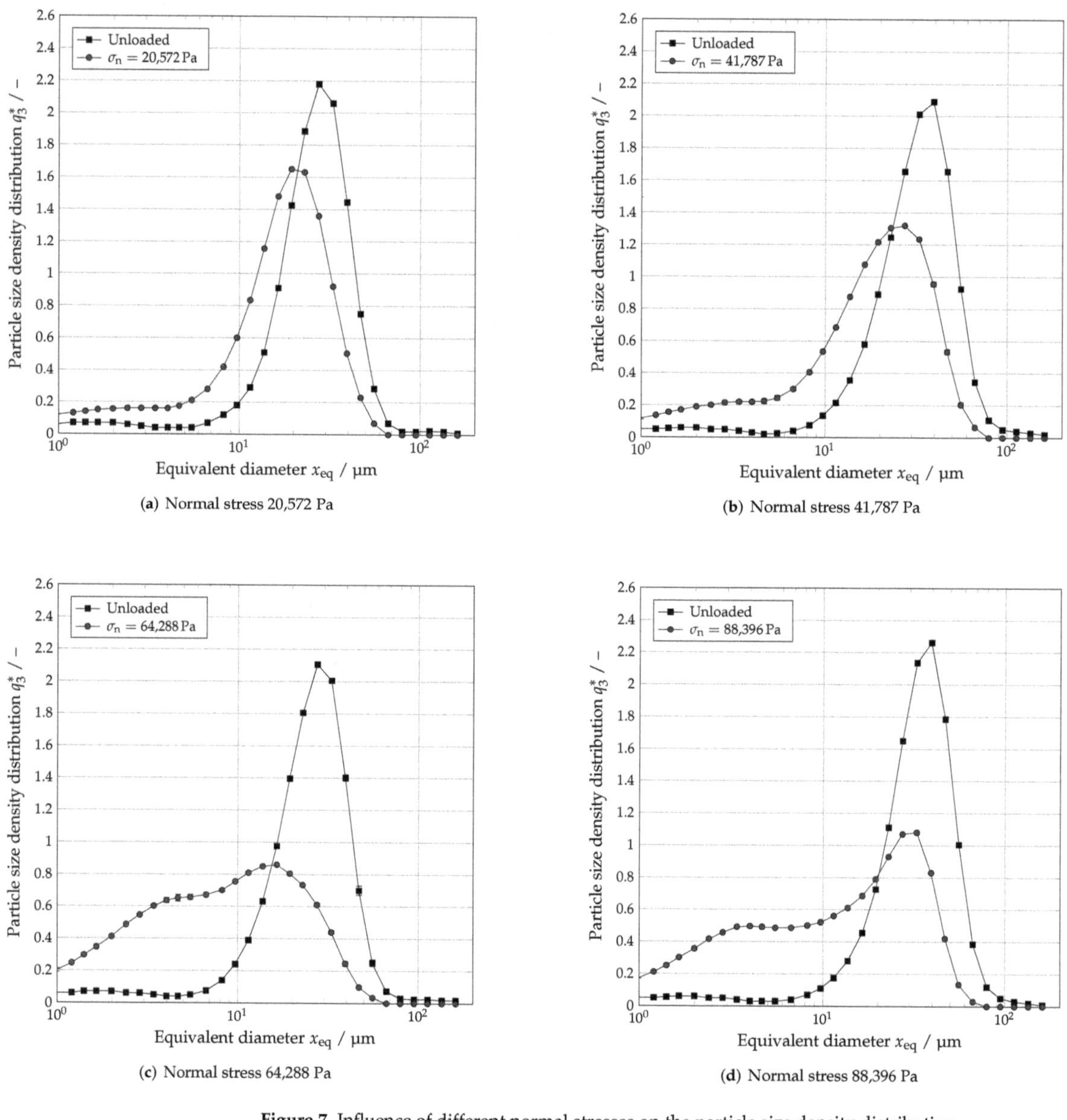

(**a**) Normal stress 20,572 Pa

(**b**) Normal stress 41,787 Pa

(**c**) Normal stress 64,288 Pa

(**d**) Normal stress 88,396 Pa

Figure 7. Influence of different normal stresses on the particle size density distribution.

At a higher normal stress of 41,787 Pa, the same effects occur more strongly. The distribution width increases by 62% compared to the unloaded specimen. The characteristic diameters x_{90} and x_{50} are now only 74 and 58% of the initial sizes, respectively. The x_{10} diameter also decreases significantly to 19%. The density distribution shows an increase in fines and a reduction of large particles. Compared to the normal stress of 20,572 Pa, the comminution at 41,787 Pa is more pronounced and the PSD is significantly wider. In the density distribution, this fact is shown by the reduction and broadening of the peak.

A further increase of the normal stress to 64,288 Pa leads to an obvious change in the particle size density distribution. The peak, still clearly visible in the unloaded sample, has almost completely disappeared. A strong increase of very small particles and an intense broadening of the distribution can be observed. The amount of larger particles decreases somewhat, but is still present in the shear-loaded sample. The distribution width is 147% higher compared to the unloaded sample. The x_{90} diameter decreases to 63%. The median diameter also decreases significantly and is 32% of the initial diameter. This is a much greater decrease compared to the lower normal loadings. The x_{10} diameter amounts to 15% of the initial unloaded diameter.

An increase of the normal stress beyond this to 88,396 Pa does not cause any further significant comminution of the particles. The distribution width is broadened with an increase of 136%. This is similar to the previous normal stress. The x_{90} diameter is 68% of the initial diameter. The median and x_{10} diameters are reduced to 38 and 12%, respectively. These values are also similar to the previous normal stress level. There is a clear increase in small particles.

The comparison of the different normal stresses shows a clear influence of the applied stress on the occurring comminution. Low normal stresses result in less comminution. However, even small normal stresses cause a widening of the PSD and a significant decrease of the x_{10} diameter. With an increase of the normal stress, the comminution intensifies and the distribution width increases significantly. The median diameter is strongly reduced. Since the shear load in the sediment has a gradient, large particles are always retained, which is reflected in a less pronounced reduction of the x_{90} diameter. Above a normal stress of 64,288 Pa, an increase in normal stress does not result in additional comminution. The decrease in characteristic diameter, increase in distribution width, and comminution are similar. This indicates an upper limit above which the normal stress loses its influence. The comminution is probably not only due to particle breakage but also to abrasion. Abrasion is caused by the friction of the crystals against each other, which, for example, abrades the corners of the crystals.

4.3. *Influence of Shear Velocity*

In addition to the normal stress, an influence of the shear velocity on the comminution is also conceivable. In order to investigate this influence, sediments of isometric lysozyme crystals are loaded with the three shear velocities 0.48; 1.5 and 4.5 mm min^{-1} at a constant normal stress of 64,288 Pa and a constant shear path. The particle size density distributions of the unloaded and sheared specimens with different shear velocities are shown in Figure 8.

Even at the lowest shear velocity in Figure 8, top left, there is a strong comminution of the particle collective. The distribution width increases by 138% and the x_{90} and median diameters decrease to 75 and 41%, respectively. The x_{10} diameter reduces significantly to 15% compared to the unloaded suspension. Increasing the shear velocity to 1.5 mm min^{-1} shows higher comminution for the x_{90} and x_{50} diameters and an increase in the distribution width. The respective values are analogous to Figure 7 at the corresponding normal load of 64,288 Pa. A further increase of the shear velocity to 4.5 mm s^{-1} leads to a very similar particle size density distribution of the sheared crystals compared to the previous shear velocity. The reduction of the characteristic median and x_{90} diameter is more pronounced by a few percentage points.

The lower right plot in Figure 8 shows the particle size density distribution of a sediment after shearing at a speed of 4.5 mm min^{-1} and a normal stress of σ_n = 20,572 Pa. Here, it can be seen that the normal stress has a stronger influence on the comminution than the shear velocity. At a high shear velocity and low normal stress, the comminution is less. The x_{90} diameter is still 80% of the initial particle size. The median diameter decreases to 69% and the x_{10} diameter reduces significantly to 15%. The reduction in the median and x_{90} diameters thus corresponds almost exactly to the values at a normal load of 20,572 Pa and a shear velocity of 1.5 mm min^{-1} from Figure 7. However, the reduction in x_{10} diameter is more pronounced at the higher shear velocity.

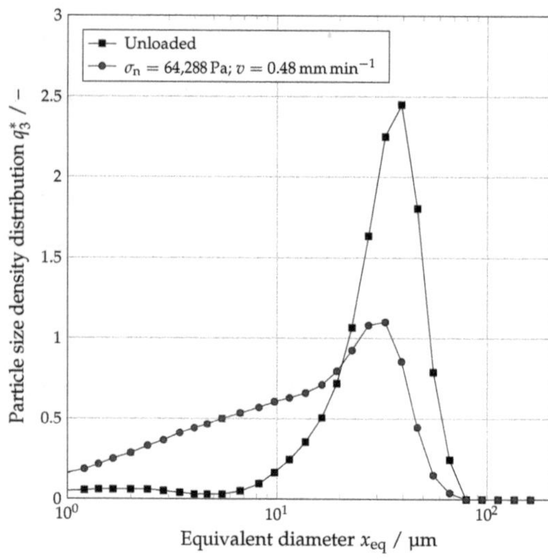

(a) Normal stress 64,288 Pa; shear velocity 0.48 mm min^{-1}

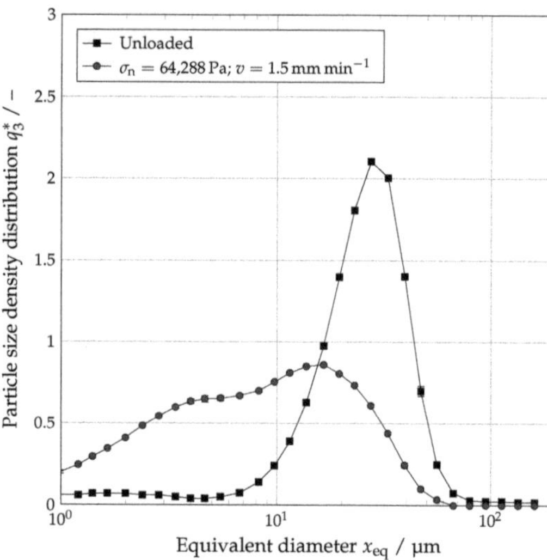

(b) Normal stress 64,288 Pa; shear velocity 1.5 mm min^{-1}

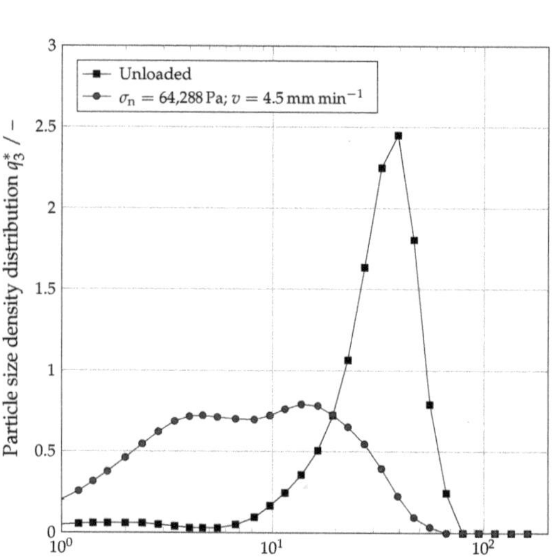

(c) Normal stress 64,288 Pa; shear velocity 4.5 mm min^{-1}

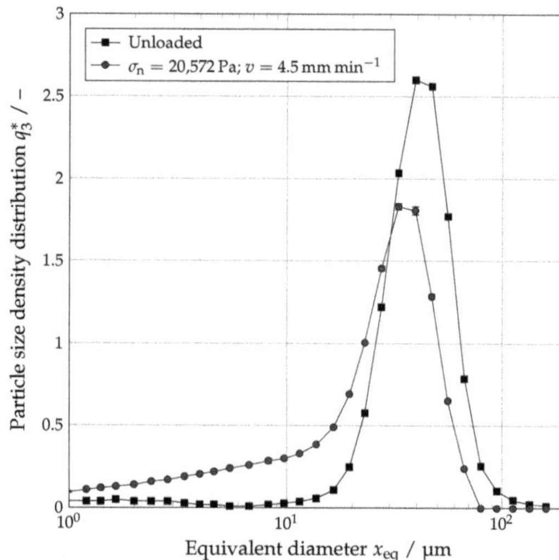

(d) Normal stress 20,572 Pa; shear velocity 4.5 mm min^{-1}

Figure 8. Influence of different shear velocities on the particle size density distribution.

5. Conclusions

In summary, it can be stated that crystal breakage always occurs in the case of combined normal and shear loading of lysozyme crystals. This is also true for normal stresses, for which no reduction in particle sizes can be detected without shear loading. In the case of combined stress, an increase in normal stress until a value of 64,288 Pa leads to significantly higher comminution. The shear velocity, on the other hand, has a much smaller effect on comminution and is a negligible influence compared to the normal stress. The high increase in fines can be explained by crystal abrasion in addition to crystal breakage. A higher normal stress increases the friction in the sediment, which is why crystal corners and edges rub off. This contributes to the strong increase in fines.

Superposition of compressive and shear stress occurs in many technical apparatuses and must, therefore, be minimized in order to avoid undesirable comminution of protein crystals. Examples of an apparatus with superimposed stress include pumps, tubular centrifuges, decanter centrifuges, and cross-flow filters. With high crystal concentrations, such as those that occur during water evaporation crystallization, the risk of particle abrasion or particle collision is increased.

Author Contributions: Conceptualization, B.R. and H.N.; methodology, B.R.; validation, B.R.; formal analysis, B.R.; investigation, B.R.; resources, H.N.; data curation, B.R.; writing—original draft preparation, B.R.; writing—review and editing, B.R., M.G. and H.N.; visualization, B.R.; supervision, M.G. and H.N.; project administration, B.R. and H.N.; funding acquisition, H.N. All authors have read and agreed to the published version of the manuscript.

Funding: This research was funded by Deutsche Forschungsgemeinschaft grant number NI 414/26-2. We acknowledge support by the KIT-Publication Fund of the Karlsruhe Institute of Technology.

Institutional Review Board Statement: Not applicable.

Informed Consent Statement: Not applicable.

Data Availability Statement: The data presented in this study are available on request from the corresponding author.

Conflicts of Interest: The authors declare no conflict of interest.

Abbreviations

The following abbreviations are used in this manuscript:

µCT	Micro Computed Tomography
PSD	Particle Size Distribution

References

1. Hubbuch, J.; Kind, M.; Nirschl, H. Preparative Protein Crystallization. *Chem. Eng. Technol.* **2019**, *42*, 2275–2281. [CrossRef]
2. Basu, S.K.; Govardhan, C.P.; Jung, C.W.; Margolin, A.L. Protein crystals for the delivery of biopharmaceuticals. *Expert Opin. Biol. Ther.* **2004**, *4*, 301–317. [CrossRef] [PubMed]
3. Hekmat, D. Large-scale crystallization of proteins for purification and formulation. *Bioprocess Biosyst. Eng.* **2015**, *38*, 1209–1231. [CrossRef] [PubMed]
4. Hallas-Møller, K.; Petersen, K.; Schlichtkrull, J. Crystalline and amorphous insulin-zinc compounds with prolonged action. *Science* **1952**, *116*, 394–398. [CrossRef] [PubMed]
5. Groß, M.; Kind, M. Bulk Crystallization of Proteins by Low-Pressure Water Evaporation. *Chem. Eng. Technol.* **2016**, *39*, 1483–1489. [CrossRef]
6. Radel, B.; Funck, M.; Nguyen, T.H.; Nirschl, H. Determination of filtration and consolidation properties of protein crystal suspensions using analytical photocentrifuges with low volume samples. *Chem. Eng. Sci.* **2019**, *196*, 72–81. [CrossRef]
7. Radel, B.; Nguyen, T.H.; Nirschl, H. Calculation of the flux density function for protein crystals from small scale settling and filtration experiments. *AIChE J.* **2021**, *67*, e17378. [CrossRef]
8. Dobler, T.; Radel, B.; Gleiss, M.; Nirschl, H. Quasi-Continuous Production and Separation of Lysozyme Crystals on an Integrated Laboratory Plant. *Crystals* **2021**, *11*, 713. [CrossRef]
9. Barros Groß, M.; Kind, M. Comparative Study on Seeded and Unseeded Bulk Evaporative Batch Crystallization of Tetragonal Lysozyme. *Cryst. Growth Des.* **2017**, *17*, 3491–3501. [CrossRef]
10. Walsh, G. *Proteins*, 2nd ed.; John Wiley & Sons Inc.: Hoboken, NJ, USA, 2014.
11. Cornehl, B.; Overbeck, A.; Schwab, A.; Büser, J.P.; Kwade, A.; Nirschl, H. Breakage of lysozyme crystals due to compressive stresses during cake filtration. *Chem. Eng. Sci.* **2014**, *111*, 324–334. [CrossRef]
12. Cornehl, B.; Grünke, T.; Nirschl, H. Mechanical Stress on Lysozyme Crystals during Dynamic Cross-Flow Filtration. *Chem. Eng. Technol.* **2013**, *36*, 1665–1674. [CrossRef]
13. Jen, A.; Merkle, H.P. Diamonds in the rough. *Pharm. Res.* **2001**, *18*, 1483–1488. [CrossRef] [PubMed]
14. Illies, S.; Pfinder, J.; Anlauf, H.; Nirschl, H. Filter cake compaction by oscillatory shear. *Dry. Technol.* **2016**, *35*, 66–75. [CrossRef]
15. Höfgen, E.; Collini, D.; Batterham, R.J.; Scales, P.J.; Stickland, A.D. High pressure dewatering rolls: Comparison of a novel prototype to existing industrial technology. *Chem. Eng. Sci.* **2019**, *205*, 106–120. [CrossRef]

16. Hammerich, S.; Stickland, A.D.; Radel, B.; Gleiss, M.; Nirschl, H. Modified shear cell for characterization of the rheological behavior of particulate networks under compression. *Particuology* **2020**, *51*, 1–9. [CrossRef]
17. Tiller, F.M.; Lu, W.M. The role of porosity in filtration VIII: Cake nonuniformity in compression–permeability cells. *AIChE J.* **1972**, *18*, 569–572. [CrossRef]
18. Schönert, K. The influence of particle bed configurations and confinements on particle breakage. *Int. J. Miner. Process.* **1996**, *44–45*, 1–16. [CrossRef]
19. Liu, J.; Schönert, K. Modelling of interparticle breakage. *Int. J. Miner. Process.* **1996**, *44–45*, 101–115. [CrossRef]
20. Rumpf, H. Grundlegende physikalische Probleme bei der Zerkleinerung. *Chem. Ing. Tech.* **1962**, *34*, 731–741. [CrossRef]

Article

Structure-Based Modeling of the Mechanical Behavior of Cross-Linked Enzyme Crystals

Marta Kubiak *, Ingo Kampen and Carsten Schilde

Institute for Particle Technology, Technische Universität Braunschweig, Volkmaroder Str. 5, 38104 Braunschweig, Germany; i.kampen@tu-bs.de (I.K.); c.schilde@tu-bs.de (C.S.)
* Correspondence: marta.kubiak@tu-braunschweig.de; Tel.: +49-(0)53139165533

Abstract: Because of their high volumetric catalytic activity, in addition to their high chemical and thermal resistances, enzymes in the form of protein crystals are an excellent choice for application as immobilized biocatalysts. However, mechanical stability is a requirement for the processability of immobilisates, in addition to the protein crystals retaining their enzymatic activity, and this is closely related to the crystal structure. In this study, the influence of protein engineering on the mechanical stability of cross-linked enzyme crystals (CLECs) was investigated using a genetically modified model protein in which additionally cysteines were introduced on the protein surface for targeted cross-linking. The results showed that the mechanical stability of crystals of the mutant proteins in the native form was decreased compared to native wild-type crystals. However, specific cross-linking of the introduced amino acid residues in the mutant proteins resulted in their increased mechanical stability compared to wild-type CLECs. In order to determine the correlation between the crystal structure and the resulting mechanical properties of CLECs to enable targeted cross-linking, a previously developed model was revised and then used for the two model proteins. This model can explain the mechanically investigated relationships, such as the anisotropic crystal behavior and the influence of a linker or mutation on the micromechanical properties and, hence, can be helpful for the tailor-made production of CLECs.

Keywords: protein engineering; CLECs; modeling

Citation: Kubiak, M.; Kampen, I.; Schilde, C. Structure-Based Modeling of the Mechanical Behavior of Cross-Linked Enzyme Crystals. *Crystals* **2022**, *12*, 441. https://doi.org/10.3390/cryst12040441

Academic Editor: Abel Moreno

Received: 1 March 2022
Accepted: 19 March 2022
Published: 22 March 2022

1. Introduction

In comparison to chemical synthesis, enzyme-catalyzed reactions proceed with high chemo-, regio-, and enantioselectivity [1,2]. Because of their high efficiency, environmental friendliness, and safety [3,4], they are widely used in numerous industrial sectors, such as pharmaceutical research, food modification, biofuel development, agroindustry, and laundry [5]. The great application potential of enzymes has led to huge growth in their consumption, which is why the global market for enzymes was estimated at $9.9 billion in 2019 and is expected to grow with an annual average growth rate of 7.1% from 2020 to 2027 [6].

However, the advantages are offset by the high requirements of the enzymes with regard to the very defined and mild reaction conditions, such as temperature, pressure, or solvents used [7,8]. Their technical use is also limited by low long-term stability, as well as difficult processability and reusability [9]. Therefore, to ensure the efficient application of enzymes in biocatalysis, some optimization in terms of stability, activity, or tolerance to the solvents is necessary [10]. Originally, in order to increase the efficiency of the biocatalysts, the reaction conditions of synthesis were adapted to the enzyme requirements, leading to the development of the so-called medium engineering approach [11–13]. In the meantime, the focus is shifting to the development of other strategies for improving the performance of biocatalysts. Two possible concepts that can effectively contribute to the improvement of enzyme properties are protein engineering and enzyme immobilization. Protein engineering aims to improve enzyme properties, such as activity or selectivity, through selective manipulation of amino acids. In immobilization, a soluble enzyme is physically separated or localized in a confined space while retaining its catalytic activity [14]. This can result in an increase in productivity and enzyme lifetimes, as well as improvement in handling and facilitation of recovery [15]. The simultaneous use of both strategies led to the emergence of a new field of research called immobilized biocatalyst engineering in which the aim is to produce tailor-made immobilized biocatalysts for use in bioprocesses [16]. One of the most convincing concepts for the stabilization and immobilization of enzymes is to cross-link enzymes in their crystalline state. Such cross-linked enzyme crystals (CLECs) remain active under comparatively extreme conditions of temperature, pH, or organic solvent content [17]. In addition, CLECs show high resistance against autolysis and degradation by proteases and show good storage stability at room temperatures [18].

However, the production of efficient, tailor-made CLECs is limited by the lack of understanding of the relationship between an enzyme's structure and the resulting properties of the crystals built from that enzyme. In addition to the enzyme itself, the mechanical stability of these protein crystals is closely related to the crystal structure. The challenge of protein engineering lies in the prediction of potential amino acid sequence changes in order to obtain suitable protein structures that then fulfill the specified functions. Due to the complex and individual behavior of enzymes, it is extremely difficult to predict the overall effect of a sequence change on the structure and, thus, on the function. To be able to produce tailor-made CLECs, all the interactions, limitations, and influences along the entire production chain, starting from the sequence to the product, must therefore be investigated and elucidated.

In this study, the influence of protein engineering on the mechanical stability of CLECs was investigated and modeled based on a model protein, namely, halohydrin dehalogenase (HheG), which is used, for example, to remove haloalcohols in food applications [19]. As a basis, amino acids on the surface of the folded wild-type enzyme structure were genetically exchanged using protein engineering methods, as described in Staar et al. [17], to incorporate new potential cross-linking sites within the subsequently fabricated protein crystals and thus improve the mechanical crystal properties. The focus was on elucidating the structure–property and sequence–property relationships based on the performed experiments. For this purpose, a previously developed model [20] was expanded and then validated using the new crystals presented here, as well as for various cross-linked wild-type crystals. The results of our study establish the fundamentals for the desired formulation of reinforced catalytic active crystalline biocatalysts and, therefore, their expanded applications in industrial processes.

2. Materials and Methods

2.1. Experimental Procedure

In order to reliably measure the mechanical properties, certain requirements with regard to crystallization must be met: The crystals must (i) adhere to the surface of the sample holder, (ii) be present in individual units, (iii) the crystal shape must be fully formed, and (iv) the crystal face should be placed orthogonal to the direction of measurement. Furthermore, crystals must (v) have a minimum particle size (approximately 70 μm) and (vi) exhibit reproducible production. Analogous to crystallization, the subsequent cross-linking also had to meet certain criteria, such as (i) complete and homogeneous cross-linking within the crystal, (ii) no negative influence on the catalytic activity, (iii) no change in the surface properties of the crystals, (iv) no crystal damage after cross-linking, and (v) no detachment of the crystals from the surface of the sample carrier. Figure 1 shows a sample of cross-linked HheG crystals that completely fulfills these requirements.

Figure 1. Cross-linked hexagonal HheG wild-type crystals.

In order to examine the structure–property relationships of CLECs, the model protein halohydrin dehalogenase HheG from *Ilumatobacter coccineus* (HheG) was produced and purified by the working group of Prof. A. Schallmey from the Institute for Biochemistry in Braunschweig and provided for the experimental study. In addition, through mutagenesis of the wild-type HheG proteins, aspartic acid was exchanged for cysteine at position 114 to insert a new cross-linking site for BMOE (bismaleimidoethane) on the surface of the protein, as previously described [17].

An AFM (Bruker NanoWizard3, former JPK) and a classical nanoindenter (Hysitron TriboIndenter Ti900) were used to characterize the mechanical properties at different scales. The AFM allowed for the mechanical properties of the surface at low forces to be investigated by generating small penetration depths in the nanometer range (compare with [20,21]). The nanoindenter allowed for the use of much higher indentation forces such that mechanical behavior at large penetration depths can be investigated. The mechanical investigation at both depth ranges was of particular interest from a modeling view, which was based on a molecular crystal structure. With just a few molecules, the number of data points that needed to be processed was already in the seven-digit range. This posed a major computational problem due to the limitations of computational capacity. Hence, the modeling had to be limited to a relatively small crystal section. To verify whether the model reproduced the mechanical properties within the entire crystal, depth-dependent mechanical measurements were performed using an AFM and a nanoindenter.

2.2. Crystallization and Cross-Linking

Wild-type halohydrin dehalogenase HheG and mutated HheG proteins (D114C) were crystallized using a sitting drop method such that the crystals were grown on a siliconized slide and, hence, they adhered to the surface (compare Section 2.1). A 20 µL droplet composed of protein stock solution (32 mg/mL for wild-type HheG and 24 mg/mL for D114C) and precipitation solution (PEG 4000 (10% (*w*/*v*) in HEPES buffer (10 mM, pH 7.3)) was placed on a cover slide and equilibrated against reservoir solution (500 µL) at 5 °C for a few days.

The procedure for cross-linking followed already established rules [20]. First, the mother liquor was removed to get rid of any remaining free proteins in the droplet, which would otherwise also be cross-linked and form an undefined precipitate. For this purpose, the sample with the crystals had to be placed on ice for about one hour to reduce the osmotic shock during the solution change and avoid crystal breakage [22]. The crystals were subsequently washed several times by adding about 10 μL of the crystallization solution to the crystals and carefully wiping it off with precision wipes without touching the crystals. Depending on the cross-linking method, the linker was added either directly onto the crystals (soaking method) or into a reservoir so that the crystals were cross-linked via the vapor diffusion and hanging drop method. Five cross-linkers having different lengths and properties, as shown in Table 1, were applied for the cross-linking of the model protein crystals.

Glutaraldehyde solution 50% (Sigma-Aldrich Chemie GmbH, Taufkirchen, Germany) is the most commonly used linker for cross-linking protein crystals. DMP (dimethyl pimelimidate), DST (disuccinimidyl tartrate), and Sulfo-EGS (ethylene glycol bis(sulfosuccinimidyl succinate)), purchased from Thermo Fisher Scientific, are specific lysine linkers with a well-defined length. These linkers were used for cross-linking wild-type HheG crystals. BMOE purchased from Thermo Fisher Scientific is a cysteine linker and was applied for the D114C crystals, where a cysteine residue was introduced by protein engineering.

Table 1. Length and conjugation of the used linkers.

Cross-Linker	Length (Å)	Conjugation
Glutaraldehyde (GA)	~5 and longer	Amines (lysine, arginine)
DMP	~9	Amines (lysine)
DST	~6	Amines (lysine)
Sulfo-EGS	~16	Amines (lysine)
BMOE	~8	Sulfhydryls (cysteine)

For the AFM-based nanoindentation, the wild-type HheG crystals were cross-linked using a soaking method with either a glutaraldehyde solution (5% (v/v)) or a mix of three different linkers (DMP, DST, and Sulfo-EGS), according to Table 2. For the mechanical investigation using a nanoindenter, both the wild-type and D114C HheG crystals were cross-linked with glutaraldehyde using a vapor diffusion method (25% (v/v)), as previously described [22]. After that, some of the samples of the D114C crystals were additionally cross-linked using a soaking method and the BMOE cross-linker. For mechanical quantification of cysteine cross-linking, a BMOE single cross-linker was also used for the cross-linking of D114C crystals. Table 2 summarizes the cross-linking conditions. According to Kubiak et al., 24 h is needed for complete cross-linking of HheG crystals. For that reason, all the samples were always cross-linked over 24 h.

Table 2. Stock solutions for the cross-linking. The solutions of BMOE and DST were diluted with a mixture of TE and HEPES buffer to a final concentration of 2 and 4 mM, respectively.

	Glutaraldehyde	BMOE	DST	DMP	Sulfo-EGS
Concentration	5% or 25%	20 mM	40 mM	4 mM	4 mM
Solvent	HEPES buffer	DMSO	DMSO	TE and HEPES buffer	

2.3. *Mechanical Measurements*

Methods for mechanical characterization of protein crystals were already established and are described in detail in previous publications for AFM-based nanoindentation [20,21,23] and classical nanoindentation using a nanoindenter [22]. As previously reported, CLECs were mechanically investigated in a liquid environment. For the AFM measurements, a spherical cantilever with a radius of 150 nm and a nominal spring constant of 40 N/m was used. The nanoindentation was performed with a Berkovich tip at 5 °C in a displacement-controlled mode. Further information about, e.g., detailed measurement parameters, can be found in the publications of Kubiak et al. [20,21,23].

2.4. Modeling of Anisotropic Crystal Strength

Based on the crystal structure, a structural model was developed to represent the mechanical behavior of cross-linked crystals at the structural level. MATLAB version R2017a software was used for this purpose. The procedure for modeling the mechanical behavior of the crystals was described in detail in the publication of Kubiak et al. [20] and is repeated with permission in this publication. The developed toolbox allows for reading the Protein Data Bank file format (.pdb), which represents the structure of three-dimensional molecules based on the atomic coordinates of amino acids of the protein. A .pdb file format contains coordinates from the smallest unit of the crystal (the so-called asymmetric unit), which can be multiplied to a larger crystal structure according to the given symmetry operators, which are also present in the file. The developed model allows for rebuilding the whole crystal using the symmetry operators, searching for atoms of relevant amino acids residues, e.g., sulfur atoms of the cysteine residues, and calculating the distances between them (cf. Figure 2). The amino acid residues that would be cross-linked by glutaraldehyde were determined based on the literature [24–27]. According to those studies, cross-linking bonds can be expected between three residual pairs: the ε-amines of lysine residues (Lys–Lys), two neighbored arginine residues (Arg–Arg), and arginine and lysine residues (Lys–Arg) having a maximal distance of 10 Å to each other (cf. [20]). For alternative cross-linkers (e.g., BMOE) the distance and the amino acids residues were adapted according to Table 1. For the crystals, all bonds within a unit cell and its surrounding of 100 Å (50 asymmetric units) were considered.

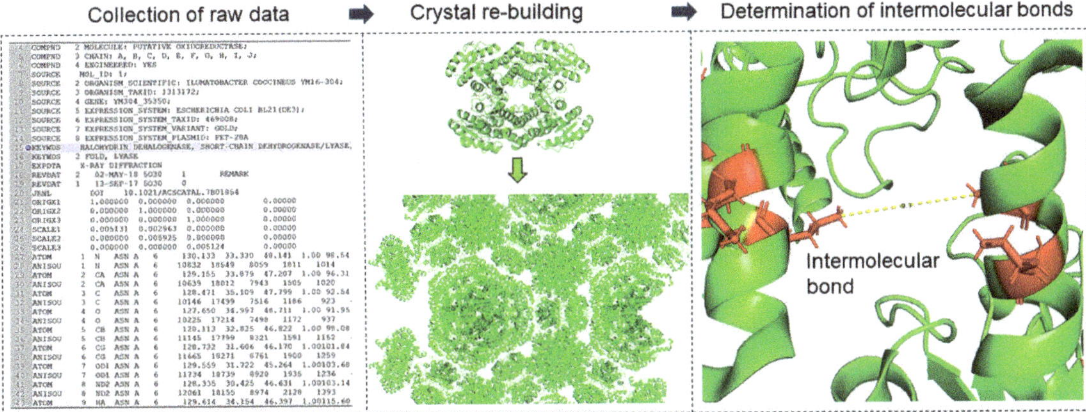

Figure 2. Overview approach to modeling mechanical crystal behavior based on crystal structure: investigation and classification of inter- and intramolecular cross-linking bonds.

Then, possible cross-linking bonds were analyzed regarding their distance and direction. Finally, the directions of the bonds in relation to the mechanically stressed crystal face were considered in order to calculate the direction-dependent crystal strength. For the investigation of the anisotropic crystal strength, the intra- and intermolecular bonds were considered separately, assuming that only the intermolecular bonds contribute to the improvement of the stability of the crystal structure in an environment other than the crystallization cocktail [20]. Furthermore, the strength of these bonds depends on their orientation with respect to the applied force during indentation, according to Figure 3.

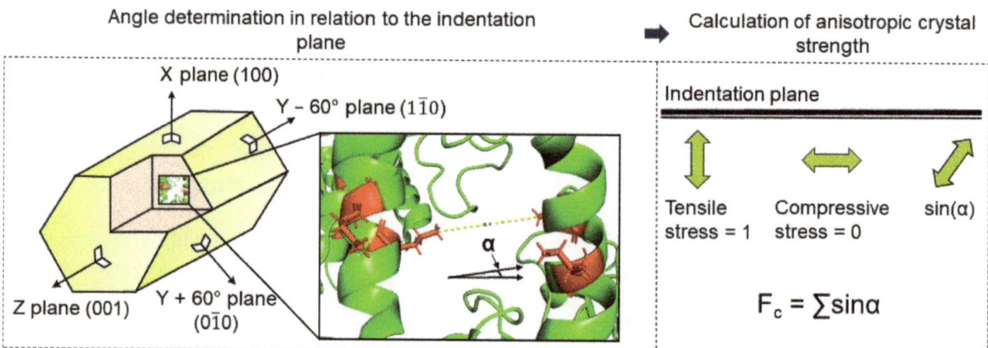

Figure 3. Overview approach to modeling mechanical crystal behavior based on crystal structure: calculation of anisotropic crystal strength using a direction of cross-linking bonds in relation to the indentation plane.

Similar studies on anisotropic particle strength as a function of porosity, particle structure, or size were published in the current literature [28–30]. Rumpf investigated a continuum-based model for determining the tensile strength of aggregates. It was based on assessing all partial stress contributions in a cross-sectional area of the structure in the direction of the applied force [28]. Schilde et al. reported that this model can also be applied to the compressive strength of aggregates measured via nanoindentation [29–31]. Based on these results, the fraction of each bond in the loading direction was calculated for each loaded crystal face. For this purpose, the sine of the angle between the normal of the crystal surface (⊥) and the direction of the bond was used as a factor. The determination of the directional crystal faces (the planes), which were used as the surface normal in the calculation, was achieved using the geometry data of the unit cell and the plane equations. As long as the function of bond strengths in terms of type and length was unknown, they were assumed to be equal and length-independent. Under this assumption, each bond was mapped to all the normal vectors of the crystal surface, as shown in Figure 3. In this way, a representative value for the total strength of the crystal surface was obtained. All the bonds were then summed proportionally to each face (resistance of the cross-linking bridges to the normal force).

3. Results and Discussion

3.1. Mechanical Properties of Wild-Type HheG Crystals in the Nanometer Range

Due to the increasing demands on the tailored properties of CLECs in terms of enzymatic activity, as well as mechanical stability, the elucidation of the relationships between the structure and the resulting properties is of particular interest. One of the most widely used linkers in the design of CLECs is glutaraldehyde [32–34]. However, according to the manufacturer's instructions, highly concentrated glutaraldehyde can be an irritant, toxic, or carcinogenic. In addition, reports indicate that excessive cross-linking with glutaraldehyde can lead to a reduction in catalytic activity [35]. For these reasons, the motivation to search for alternative cross-linkers increases. Previous studies have reported on the micromechanical properties of wild-type HheG CLECs [20], as well as the dominating influence of cross-linking bonds on the mechanical behavior of enzyme crystals [22]. The present study focused on modeling the properties of cross-linked enzyme crystals such that the selection of linkers and their influence on the mechanical and, if necessary, catalytic properties can be explained and predicted. Figure 4 compares the mechanical properties of wild-type crystals cross-linked with different linkers. Alternatively, for the glutaraldehyde, a mixture of three lysine linkers of defined length (6, 9, and 16 Å) was used for cross-linking. A single linker would not be sufficient to adequately cross-link the crystals; therefore, a mixture of linkers of different lengths had to be used.

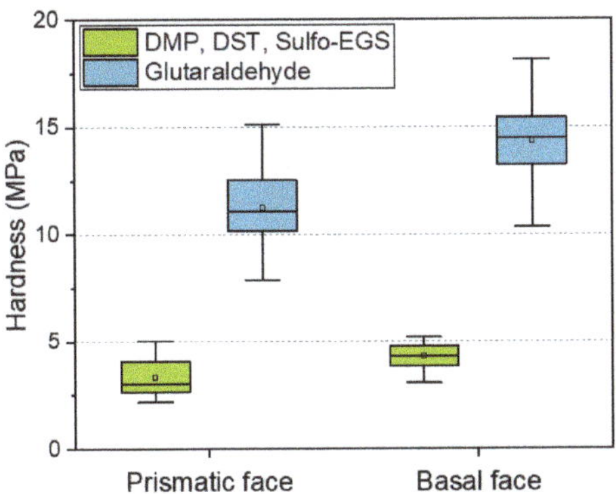

Figure 4. Comparison of the hardness of wild-type crystals cross-linked using the different linkers for 24 h. Results of WT HheG, cross-linked with glutaraldehyde, were adapted with permission from ref. [20]. Copyright © 2019, American Chemical Society.

Because glutaraldehyde tends to self-polymerize, the resulting cross-linking bond lengths are unpredictable [36]. In addition, excessive cross-linking can lead to a decrease in enzymatic activity [35]. For controllable and predictable cross-linking treatment, alternative linkers can be applied. However, our results showed that, using the alternative lysine cross-linker, the hardness of the anisotropic crystal faces decreased by about 70% compared to the cross-linking with the glutaraldehyde. Besides the high stability of CLECs cross-linked with GA, its low price is of great advantage. A kilogram of 50% glutaraldehyde solution costs only about EUR 70. In comparison, alternative linkers cost a thousand times more (DMP: 159 EUR/1 g; DST: 254 EUR/50 mg; Sulfo-EGS: 185 EUR/50 mg). Predicting the mechanical behavior based on a model can therefore bring advantages in terms of financial expenditure and experimental effort. Table 3 summarizes the modeled direction-dependent crystal strength of all intermolecular bonds smaller than 10 Å (cross-linking with GA), or Lys–Lys* bonds with a well-defined distance for alternative cross-linker (6, 9, and 16 Å, compare with Table 1).

Table 3. Summary of the direction-dependent crystal strength modeled using the anisotropic intermolecular cross-linking bonds. Adapted with permission from ref. [20]. Copyright © 2019, American Chemical Society.

	Arg–Arg	Arg–Lys	Lys–Lys	Lys–Lys*
Z	271.77	142	145.4	174.68
X	231.63	137.81	107.04	133.07
Y + 60°	230.52	140.04	105.47	132.11
Y − 60°	230.49	141.84	107.97	135.75

In Figure 5, a comparison of the experimental and modeled results is shown. The experimental and modeled data were interpreted by comparing the percentage differences between the anisotropic faces. For instance, the difference in the hardness of the basal face to the prismatic face of crystals cross-linked at the same conditions (glutaraldehyde or DMP, DST, Sulfo-EGS-linker mix, referred to as Lys–Lys* in Table 3). These percentage differences are shown as a bar (experiment) in Figure 5. Analogously, the percentage difference in the modeling results, e.g., between the anisotropic crystal faces, is shown as a dot–line plot (model). Since the mechanical properties are presented as a distribution, the lower and upper quartile (25% and 75%, respectively) were added for comparison and shown as error bars.

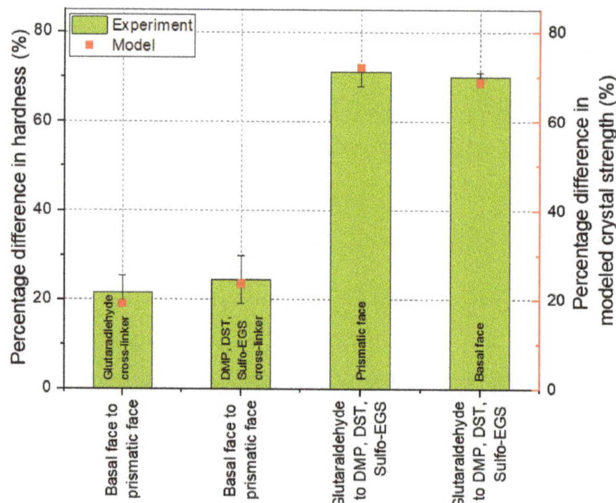

Figure 5. Comparison of the experimental and the modeled results of differently cross-linked wild-type HheG crystals. The bars represent the percentage differences between the mean hardness of the respective crystal faces considered, as described in the diagram. The square dots represent the modeling results, which are also represented as the percentage difference between the corresponding crystal faces. The error bars show the percentage differences of the upper and lower quartiles of the distribution of mechanical properties.

Originally, the model was developed by Kubiak et al. and used to explain the time-dependent anisotropy of wild-type HheG CLECs. The authors showed that crystal strength is direction-dependent due to the cross-linking bonds. The results showed that the crystal strength resulting from the Lys–Lys and Arg–Arg bonds was about 20% higher in the Z-direction (for the basal face), which is consistent with the experimental results. Looking at the cross-linking using the linker mix, it also turns out that the experiment agreed well with the model. The experimentally determined anisotropy between the basal and prismatic faces was about 24%. This corresponded to the difference in crystal strength due to the Lys–Lys bridges from the three defined linker lengths (see Table 3, Lys–Lys*, ratio of bond strength of the Z-direction (174.68) and an average of X, Y + 60°, and Y − 60° (ca. 133.64)). Previously, it was found that the Lys–Lys and Arg–Arg bridges lead to the anisotropic behavior of the HheG crystals. Assuming that there are bonds of all the considered amino acid residue pairs in the crystal, the Arg–Lys bond here did not contribute to the anisotropic behavior of the wild-type HheG crystals, and the bonds in the defined directions were summed up and the total strength of the respective areas for differently cross-linked crystals was proportionally adjusted. The mechanical investigation showed that, due to the use of a linker mix instead of glutaraldehyde, the average hardness of the prismatic and basal crystal faces decreased by about 70%. The results corresponded to the modeling with some slight deviation.

The knowledge that a crystal structure can be used to predict the mechanical behavior of the surfaces saves an enormous amount of time through the targeted selection of the linker and the subsequent model prediction of the crystal strength based on the formed cross-link bridges. The model can be used, for example, to determine which binding-site distances occur most frequently and which crystal strength will correlate with this linker length, and the length of the linker can then be accordingly selected for experimental cross-linking. In addition, it can be determined whether other linkers, e.g., carboxyl or sulfhydryl instead of amine linkers, will provide better mechanical performance. By selectively combining different variants—linker length, type, with binding sites, e.g., far away from the catalytic site—the desired mechanical strength can be modeled in advance.

3.2. Mechanical Properties of Wild-Type HheG Crystals in the Micrometer Range

In the introduction, it was written that protein engineering is a common method for improving the performance of enzymes. Unfortunately, improvements in the properties of solubilized proteins are not always accompanied by improvements in the mechanical stability of the crystals. Figure 6 shows a comparison of the fraction of elastic energy of the wild-type and the D114C crystals mechanically characterized in their native state. From the results, it can be seen that the fraction of elastic energy of the basal face of D114C crystals was significantly lower than that of the wild-type crystals. The difference was about 40% at a 200 nm penetration depth and even almost 60% at a 900 nm penetration depth. In comparison, the difference between prismatic crystal faces was much smaller and lay within the spread of the distribution of mechanical properties.

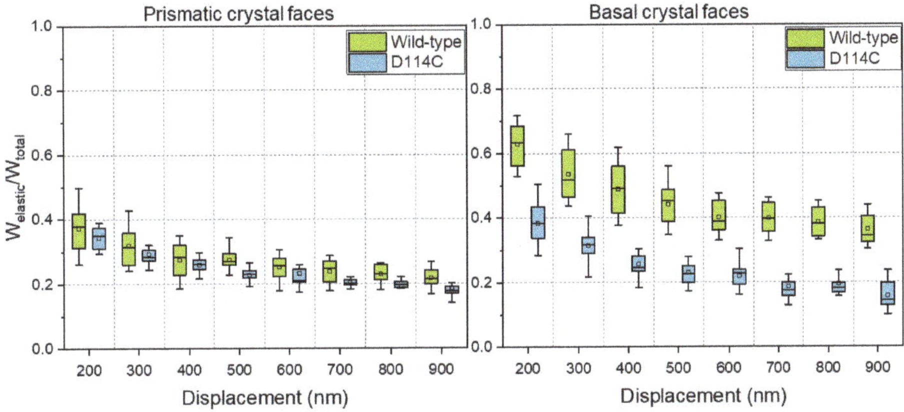

Figure 6. Influence of protein engineering on the fraction of elastic energy of the HheG enzyme crystals.

Most likely, the lowering of the elastic energy fraction of the basal crystal face was due to the point mutation of the crystal contact, which lay exactly along the *c*-axis (Z-direction), as shown in Figure 7. Since the crystals were held together in their three-dimensional form by crystal contacts, the mutation of such a site could contribute to the reduction of crystal strength. Kubiak et al. reported that the slip planes are aligned orthogonal to the indenter tip during nanoindentation of the basal crystal face, which is a reason for the easy and unhampered gliding of the planes [22]. For the native D114C crystals, it seems to reduce the interactions between the slip planes such that a lower fraction of elastic energy can be observed.

In a previous study, Kubiak et al. presented and discussed, in detail, the influence of cross-linking on the mechanical behavior of enzyme crystals. The authors also investigated the dominating influence of cross-linking on the mechanical behavior of CLECs [22]. In this section, the influence of the genetic modification of position D114 on the mechanical properties of CLECs is presented. By replacing it with cysteine, a cross-linking site was incorporated, which could be cross-linked by means of a suitable linker—in this case, a BMOE linker—in addition to the lysines. Hence, in the present study, crystals were partially cross-linked using glutaraldehyde and partially using a BMOE linker via the soaking method. In addition, some of the crystal samples were cross-linked first with GA and then with BMOE. All three samples were mechanically examined and compared with the cross-linked wild-type crystals.

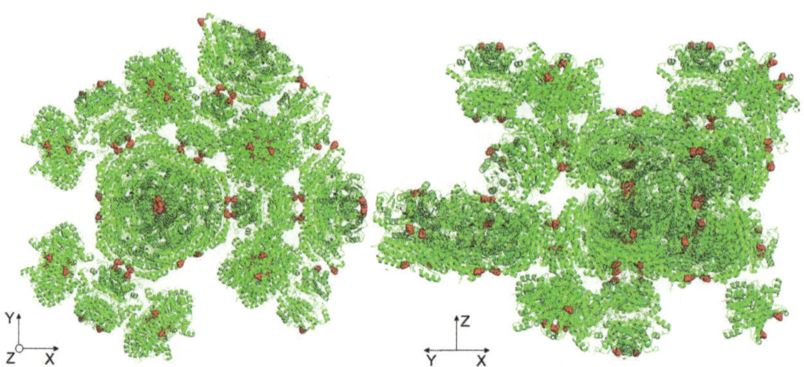

Figure 7. Highlighting the manipulated amino acids (cysteine) at one of the crystal contacts along the *c*-axis (Z-direction).

In Figure 8, the influence of the cross-linker on the elastic energy fraction of the prismatic and basal faces of D114C crystals is presented. From Figure 8, it can be seen that the fraction of elastic energy for both anisotropic faces increased with the degree of cross-linking and was the lowest for cross-linking with the BMOE linker and highest for cross-linking with both linkers. Regarding the prismatic face at the penetration depth of 200 nm, the fractions of elastic energy were 39% (BMOE), 66% (GA), and 69% (GA + BMOE). In comparison, the fraction of elastic energy of the wild-type crystals cross-linked with GA was 64%, which was slightly lower than the median value of D114C crystals that were also cross-linked with GA. Analogous to the results presented earlier about the fractions of elastic energies of native crystals, the fraction of elastic energy decreased with increasing depth. At the penetration depth of 900 nm, the fraction of elastic energy of the crystals cross-linked with BMOE was the lowest at 20% and increased to 40% when the crystals were cross-linked with GA. Due to the double cross-linking (GA + BMOE), the fraction of elastic energy was 7% higher, and at 47%, it was ca. 20% higher than the wild-type crystals. Independent of the cross-linker used, both of the cross-linked crystal faces exhibited a higher fraction of elastic energy than the native crystal faces. It is remarkable for the basal crystal faces that, in contrast to the prismatic faces at the penetration depth of 200 nm, the D114Cs had a lower fraction of elastic energy after cross-linking with GA (64%) than the wild-type crystal (72%). However, the trend was reversed in the deep crystal regions, allowing for the quantification of the stronger influence of cross-linking for the D114C CLECs. For the double cross-linked CLECs, the fraction of elastic energy was the highest and was equal to 51% at a penetration depth of 900 nm. Based on those measurements, it can be concluded that a combination of both strategies—protein engineering and enzyme immobilization—can effectively contribute to the enhancement of mechanical properties of CLECs.

In Section 3.1, the results of hardness measured using AFM at a penetration depth of less than about 30 nm for wild-type crystals were compared with the model. In this section, the model was used for correlation with other mechanical parameters, such as the fraction of elastic energy at a penetration depth of 900 nm. Thus, it was tested whether modeling from a small crystal section was sufficient for the global description of the mechanical behavior. Table 4 shows the results of the direction-dependent crystal strength for the respective pairs of amino acid residues considered for further analysis.

Without cross-linking, plastic deformation is the dominant characteristic deformation behavior of the crystals, whereas with cross-linking, elastic deformation is a decisive factor [22]. For this reason, it was assumed that the cross-linking via additives makes a significant contribution to the elastic stress state within the crystal and, thus, the structure-based modeling approach for the mechanical crystal behavior can be applied to describe the elastic crystal properties with regard to the different crystal directions. In Figure 9, the results of the nanoindenter are presented, where the anisotropy of the wild-type (WT) crystals was again related to the model based on the differences in the fraction of elastic energy. The fraction of elastic energy of the basal face of the WT CLECs was, on average, about 14% higher than for the prismatic face. Comparing the percentage difference with the summed bond strengths from the bond pairs considered, i.e., Lys–Lys, Lys–Arg, and Arg–Arg, the modeling results also showed a difference of about 14.6% in the Z-direction compared to the X, Y + 60°, and Y − 60° directions ($\sum_{Z\text{-direction}} = 559.17$ and mean of $\sum_{XY\text{-direction}} = 477.59$, cf. Table 3). Thus, based on the modeling results, it can be concluded that all bond pairs contributed to the crystal mechanics in the deep crystal regions. This would also explain why the anisotropy between different crystal faces decreased with increasing penetration depth.

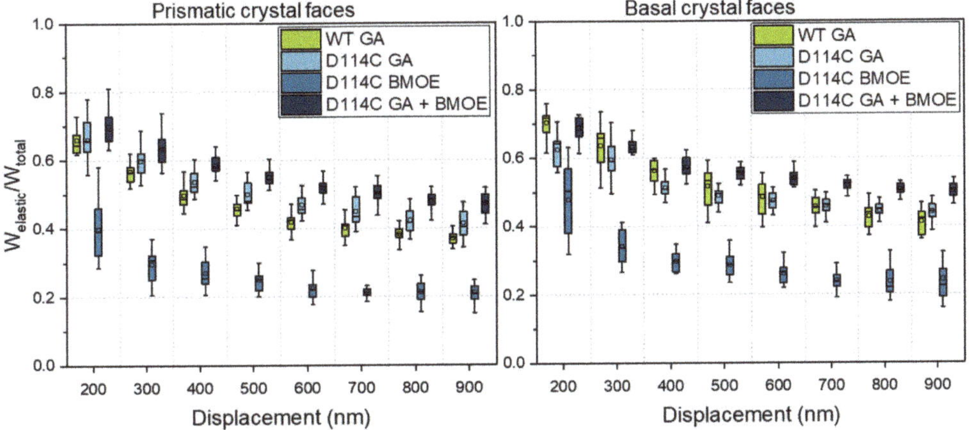

Figure 8. Fraction of elastic energy of prismatic and basal crystal faces of D114C crystals as a function of the cross-linker and penetration depth. To compare the influence of protein engineering on the mechanical behavior of CLECs, wild-type (WT) CLECs, adapted from ref. [22].

Table 4. Summary of the direction-dependent crystal strength of D114C crystals modeled using the anisotropic intermolecular cross-linking bonds.

	Arg–Arg	Arg–Lys	Lys–Lys	Cys–Cys
Z	236.30	217.39	128.73	85.32
X	214.22	236.12	125.99	82.28
Y + 60°	218.54	230.32	128.56	78.66
Y − 60°	216.19	235.65	128.15	78.54

The crystal faces of D114C mutant cross-linked by GA show about a 6% difference in the elastic energy fraction. From Table 4, it can be seen that the sum of Lys–Lys and Arg–Arg bonds (365.03 in the Z-direction and 343.88 in the X, Y + 60°, and Y − 60° directions) also showed about a 6% difference. A similar result concerned the cross-linking using GA and BMOE. Comparable to the cross-linking with GA, the difference between the model and the experimental results was negligible here. The deviation between the model and the result was the highest for cross-linking with BMOE and amounted to 30% when mean values were considered. The reason for this was that the cross-linking of cysteines alone was not sufficient to provide stable support to the crystal lattice such that increased dislocations were observed during the measurement, especially in deeper crystal regions, leading to the increased distribution width of the results. This distribution width could also be seen in the error bars. For example, a positive error bar meant that the percentage difference of the third quartile (75% of the distribution) of ca. 17% was about 90% higher than the percentage difference of the mean value (9%, see Figure 9). Nevertheless, the model result was within the error or distribution width, which demonstrated the good representation by the experiment. A comparison of the respective crystal faces (basal or prismatic face) between the two crystal structures showed that there was about a 16% deviation of the model from both the third quartile for the prismatic face and from the mean value of the distribution for the basal face.

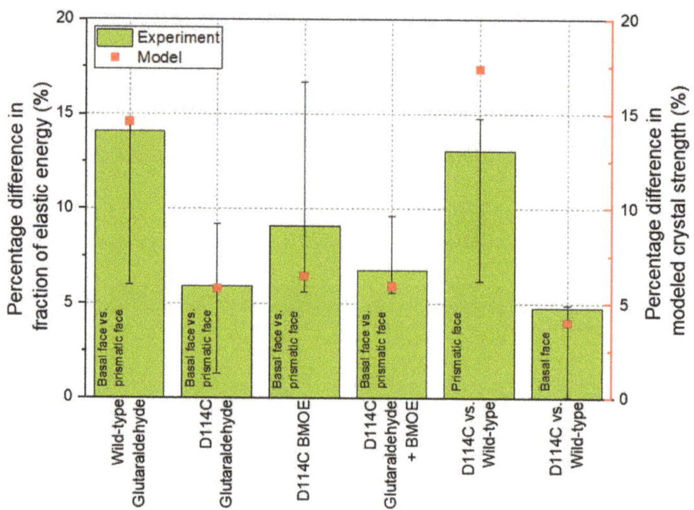

Figure 9. Comparison of the experimental and the modeled results of wild-type HheG crystals and the mutant D114C.

The most important causes, which can have a limiting influence on the modeling of the crystal behavior, can occur at three different stages of experimental execution: X-ray structure analysis, sample preparation, and mechanical measurements. For the model, the X-ray structure analysis was one of the most critical stages. Regarding crystal quality, parameters such as protein production batch or crystallization conditions and growth time may have a great influence. For the mutant D114C, the best-achieved resolution was unfortunately only 3.9 Å, which allowed for the determination of the positions of the atoms with a certain tolerance. During data processing, for example, the symmetry of the respective chains was calculated based on the crystallographic space group. Moreover, the coordinate origin was manually changed. In this process, the atoms were manually shifted to the same origin. However, the coordinates of the atoms were influenced by the B-factors and thus were not rigid but movable in the crystal and did not represent absolute positions but, rather, were relative to each other. The exact displacement measurement of the atoms was limited by the poor resolution, which is why deviations could occur. Due to the relativity of position within a crystal, this was not a critical problem when experiments and modeling were performed using the same crystal structure. However, the error may be significant if different crystal structures are compared to each other. Nevertheless, taking into account the fact that the accuracy of the crystal structures and the developed model were in the angstrom range, the agreement between the model and experiment was surprisingly high for different model proteins.

For the mechanical measurements, influencing factors such as the number of measurement points for a statistically validated estimation of the distribution of mechanical properties and the cantilever quality (geometry or durability of the measurement tip) are important. Using a careful working method, for example, by scanning the crystal surface to quantify the tilt or by regular SEM images of the cantilevers, these influences can be controlled and, if necessary, minimized.

Another serious factor that influences the modeling of the structure–property relationship is sample preparation. Depending on the formulation parameters, different crystal morphologies can be formed, resulting in differences in mechanical behavior. However, it happened that the mechanical behavior of crystals from different protein batches differed despite the crystal morphology remaining the same. This phenomenon was difficult to explain and was probably due to small differences in ions in the solution after, e.g., the purification or desalting step. Since these influencing factors are difficult to control or avoid despite using established and consistent preparation methods, sample preparation is considered a highly influential factor comparable to that of X-ray structure analysis. Nevertheless, the model reproduced the trends very well. Despite certain limitations, such as the poor resolution of the crystal structure, the model allowed for a reliable representation of the mechanical properties, both the hardness and the elastic part of the deformation energy, within the whole crystal.

The accuracy of this model could be refined by the addition of further information, e.g., regarding the bond strength as a function of bond length. Additionally, X-ray crystallographic studies of cross-linked crystals could also be performed, allowing for the resolution of systematic cross-linking bonds. Subsequently, the relevant distances/positions could be selected and modeled. It would also be possible to introduce a position-dependent strength of interactions instead of cross-linking bridges and, thus, additionally ensure the modeling of the mechanical behavior of non-cross-linked crystals. However, a prerequisite is that accurate quantifiable data on the force fields between individual molecules within the crystal are available, e.g., via molecular dynamics simulation. By combining the data from native and cross-linked crystals, the highest accuracy in terms of predicting the mechanical behavior should be achieved. However, a very high computational capacity is recommended to perform such calculations. The excerpt of crystal structures in this work was limited to 100 Å in space. Despite the relatively small crystal excerpt, the number of rows searched for targeted information in MATLAB was almost 30 million. In order to calculate the data more efficiently, the .pdb file had to be directly converted with suitable functions and limited to relevant information. In case a sensitivity analysis was to be performed using the model, all data must be available. With the help of a high-performance computer, all distances to each other could be calculated and clustered. On a qualitative basis, the amino acid positions could be located, which should be exchanged by the rational protein design to incorporate quite effective cross-linking sites. Assuming that the crystal structure will not be subject to major changes, it should therefore be possible to perform targeted mutations with this tool, which should contribute to enhancing cross-linking and mechanical properties.

4. Summary

The aim of this work was to investigate and elucidate all interactions, limitations, and influences, as well as to establish the guidelines for predicting the mechanical behavior of protein crystals. For this purpose, wild-type and genetically modified model proteins were crystallized, specifically cross-linked, and mechanically examined. Then, a mathematical model was developed, which was based on the respective crystal structure and calculated the distances and their directions between the relevant atoms of the amino acid residues. It then modeled the anisotropic total crystal strength resulting from the potential cross-linking bridges on the basis of a simple function. It was shown that this model could be used to explain all the mechanically studied relationships, such as the anisotropic crystal behavior and the influence of linkers or mutations on the micromechanical properties. By modeling the cross-linking bridges, it was possible to predict which linker length should be used to achieve the desired micromechanical properties or which amino acids should be genetically modified to incorporate positions for particularly effective cross-linking sites. It was also shown that the model could represent cross-linking behavior, as well as cross-linking tendencies resulting from protein engineering, and thus support the production of tailor-made CLECs.

Author Contributions: Conceptualization, M.K. and C.S.; Data curation, M.K.; Funding acquisition, C.S.; Investigation, M.K.; Methodology, M.K.; Software, M.K. Project administration, I.K. and C.S.; Supervision, I.K. and C.S.; Writing—original draft, M.K.; Writing—review and editing, I.K. and C.S. All authors read and agreed to the published version of the manuscript.

Funding: This work was funded by the German Research Foundation (DFG) within the priority Programme DiSPBiotech (SPP 1934, SCHI 1265/3-2).

Institutional Review Board Statement: Not applicable.

Informed Consent Statement: Not applicable.

Data Availability Statement: The data presented in this study are available within the article.

Acknowledgments: We thank Anett Schallmey and Marcel Staar (Institute for Biochemistry, TU Braunschweig) for the production and purification of the halohydrin dehalogenase proteins and Wulf Blankenfeldt and Steffi Henke for the resolution of the crystal structure of the HheG D114C crystals. We acknowledge financial support from the German Research Foundation and the Open Access Publication Funds of the Technische Universität Braunschweig.

Conflicts of Interest: The authors declare no conflict of interest.

References

1. Schmid, A.; Dordick, J.S.; Hauer, B.; Kiener, A.; Wubbolts, M.; Witholt, B. Industrial biocatalysis today and tomorrow. *Nature* **2001**, *409*, 258–268. [CrossRef] [PubMed]
2. Qu, G.; Li, A.; Acevedo-Rocha, C.G.; Sun, Z.; Reetz, M.T. The Crucial Role of Methodology Development in Directed Evolution of Selective Enzymes. *Angew. Chem. Int. Ed. Engl.* **2020**, *59*, 13204–13231. [CrossRef] [PubMed]
3. Singh, R.K.; Tiwari, M.K.; Singh, R.; Lee, J.-K. From protein engineering to immobilization: Promising strategies for the upgrade of industrial enzymes. *Int. J. Mol. Sci.* **2013**, *14*, 1232–1277. [CrossRef] [PubMed]
4. Truppo, M.D. Biocatalysis in the Pharmaceutical Industry: The Need for Speed. *ACS Med. Chem. Lett.* **2017**, *8*, 476–480. [CrossRef]
5. Bommarius, A.S.; Riebel, B.R. *Biocatalysis: Fundamentals and Applications*; Wiley-VCH: Weinheim, Germany, 2004; ISBN 3-527-30344-8.
6. Grand View Research. *Market Analysis Report: Enzymes Market Size, Share & Trends Analysis Report by Application (Industrial Enzymes, Specialty Enzymes), by Product (Carbohydrase, Proteases, Lipases), by Source, by Region, and Segment Forecasts 2020–2027*; Grand View Research: Rockville, MD, USA, 2020; pp. 1–173.
7. Buchholz, K.; Kasche, V.; Bornscheuer, U.T. *Biocatalysts and Enzyme Technology*, 2nd ed.; Wiley-Blackwell: Weinheim, Germany, 2012; ISBN 978-3527329892.
8. Brange, J.; Langkjsgmaeligr, L.; Havelund, S.; Vølund, A. Chemical stability of insulin. 1. Hydrolytic degradation during storage of pharmaceutical preparations. *Pharm. Res.* **1992**, *9*, 715–726. [CrossRef]
9. Sheldon, R.A.; van Pelt, S. Enzyme immobilisation in biocatalysis: Why, what and how. *Chem. Soc. Rev.* **2013**, *42*, 6223–6235. [CrossRef]
10. Jaeger, K.-E.; Liese, A.; Syldatk, C. *Einführung in die Enzymtechnologie*; Springer: Berlin/Heidelberg, Germany, 2018; ISBN 978-3-662-57618-2.
11. Godoy, C.A.; Fernández-Lorente, G.; de las Rivas, B.; Filice, M.; Guisan, J.M.; Palomo, J.M. Medium engineering on modified Geobacillus thermocatenulatus lipase to prepare highly active catalysts. *J. Mol. Catal. B Enzym.* **2011**, *70*, 144–148. [CrossRef]
12. Stepankova, V.; Bidmanova, S.; Koudelakova, T.; Prokop, Z.; Chaloupkova, R.; Damborsky, J. Strategies for Stabilization of Enzymes in Organic Solvents. *ACS Catal.* **2013**, *3*, 2823–2836. [CrossRef]
13. Jahangiri, E.; Agharafeie, R.; Kaiser, H.-J.; Tahmasbi, Y.; Legge, R.L.; Haghbeen, K. Medium engineering to enhance mushroom tyrosinase stability. *Biochem. Eng. J.* **2012**, *60*, 99–105. [CrossRef]
14. Katchalski-Katzir, E. Immobilized enzymes—learning from past successes and failures. *Trends Biotechnol.* **1993**, *11*, 471–478. [CrossRef]
15. Ansorge-Schumacher, M.B. Enzymimmobilisierung. In *Einführung in die Enzymtechnologie*; Jaeger, K.-E., Liese, A., Syldatk, C., Eds.; Springer: Berlin/Heidelberg, Germany, 2018; pp. 187–206. ISBN 978-3-662-57619-9.
16. Bernal, C.; Rodríguez, K.; Martínez, R. Integrating enzyme immobilization and protein engineering: An alternative path for the development of novel and improved industrial biocatalysts. *Biotechnol. Adv.* **2018**, *36*, 1470–1480. [CrossRef]
17. Staar, M.; Henke, S.; Blankenfeldt, W.; Schallmey, A. Biocatalytically active and stable cross-linked enzyme crystals of halohydrin dehalogenase HheG by protein engineering. *ChemCatChem* **2022**, e202200145. [CrossRef]
18. Navia, M.A.; Clair, N.; Griffith, J.P. Crosslinked enzyme crystals (CLECs™) as immobilized enzyme particles. *Stud. Org. Chem.* **1993**, *47*, 63–73. [CrossRef]

19. Schallmey, A. Bioinformatische Methoden zur Enzymidentifizierung. In *Einführung in die Enzymtechnologie;* Jaeger, K.-E., Liese, A., Syldatk, C., Eds.; Springer: Berlin/Heidelberg, Germany, 2018; pp. 125–140. ISBN 978-3-662-57619-9.
20. Kubiak, M.; Storm, K.-F.; Kampen, I.; Schilde, C. Relationship between Cross-Linking Reaction Time and Anisotropic Mechanical Behavior of Enzyme Crystals. *Cryst. Growth Des.* **2019**, *19*, 4453–4464. [CrossRef]
21. Kubiak, M.; Solarczek, J.; Kampen, I.; Schallmey, A.; Kwade, A.; Schilde, C. Micromechanics of Anisotropic Cross-Linked Enzyme Crystals. *Cryst. Growth Des.* **2018**, *18*, 5885–5895. [CrossRef]
22. Kubiak, M.; Staar, M.; Kampen, I.; Schallmey, A.; Schilde, C. The Depth-Dependent Mechanical Behavior of Anisotropic Native and Cross-Linked HheG Enzyme Crystals. *Crystals* **2021**, *11*, 718. [CrossRef]
23. Kubiak, M.; Mayer, J.; Kampen, I.; Schilde, C.; Biedendieck, R. Structure-Properties Correlation of Cross-Linked Penicillin G Acylase Crystals. *Crystals* **2021**, *11*, 451. [CrossRef]
24. Wine, Y.; Cohen-Hadar, N.; Freeman, A.; Frolow, F. Elucidation of the mechanism and end products of glutaraldehyde crosslinking reaction by X-ray structure analysis. *Biotechnol. Bioeng.* **2007**, *98*, 711–718. [CrossRef]
25. Yonath, A.; Sielecki, A.; Moult, J.; Podjarny, A.; Traub, W. Crystallographic studies of protein denaturation and renaturation. 1. Effects of denaturants on volume and X-ray pattern of cross-linked triclinic lysozyme crystals. *Biochemistry* **1977**, *16*, 1413–1417. [CrossRef]
26. Salem, M.; Mauguen, Y.; Prangé, T. Revisiting glutaraldehyde cross-linking: The case of the Arg-Lys intermolecular doublet. *Acta Crystallogr. Sect. F Struct. Biol. Commun.* **2010**, *66*, 225–228. [CrossRef]
27. Buch, M.; Wine, Y.; Dror, Y.; Rosenheck, S.; Lebendiker, M.; Giordano, R.; Leal, R.M.F.; Popov, A.N.; Freeman, A.; Frolow, F. Protein products obtained by site-preferred partial crosslinking in protein crystals and "liberated" by redissolution. *Biotechnol. Bioeng.* **2014**, *111*, 1296–1303. [CrossRef]
28. Rumpf, H.C.H. Zur Theorie der Zugfestigkeit von Agglomeraten bei Kraftübertragung an Kontaktpunkten. *Chem. Ing. Tech.* **1970**, *42*, 538–540. [CrossRef]
29. Schilde, C.; Burmeister, C.F.; Kwade, A. Measurement and simulation of micromechanical properties of nanostructured aggregates via nanoindentation and DEM-simulation. *Powder Technol.* **2014**, *259*, 1–13. [CrossRef]
30. Schilde, C.; Westphal, B.; Kwade, A. Effect of the primary particle morphology on the micromechanical properties of nanostructured alumina agglomerates. *J. Nanopart. Res.* **2012**, *14*, 2344. [CrossRef]
31. Schilde, C.; Kwade, A. Measurement of the micromechanical properties of nanostructured aggregates via nanoindentation. *J. Mater. Res.* **2012**, *27*, 672–684. [CrossRef]
32. Jegan Roy, J.; Emilia Abraham, T. Strategies in making cross-linked enzyme crystals. *Chem. Rev.* **2004**, *104*, 3705–3722. [CrossRef]
33. Barbosa, O.; Ortiz, C.; Berenguer-Murcia, Á.; Torres, R.; Rodrigues, R.C.; Fernandez-Lafuente, R. Glutaraldehyde in bio-catalysts design: A useful crosslinker and a versatile tool in enzyme immobilization. *RSC Adv.* **2014**, *4*, 1583–1600. [CrossRef]
34. Yan, E.-K.; Cao, H.-L.; Zhang, C.-Y.; Lu, Q.-Q.; Ye, Y.-J.; He, J.; Huang, L.-J.; Yin, D.-C. Cross-linked protein crystals by glutaraldehyde and their applications. *RSC Adv.* **2015**, *5*, 26163–26174. [CrossRef]
35. Margolin, A.L. Novel crystalline catalysts. *Trends Biotechnol.* **1996**, *14*, 223–230. [CrossRef]
36. Whipple, E.B.; Ruta, M. Structure of aqueous glutaraldehyde. *J. Org. Chem.* **1974**, *39*, 1666–1668. [CrossRef]

MDPI

Article

Stabilizing DNA–Protein Co-Crystals via Intra-Crystal Chemical Ligation of the DNA

Abigail R. Ward [1], Sara Dmytriw [2], Ananya Vajapayajula [2] and Christopher D. Snow [1,2,*]

1 Department of Chemistry, Colorado State University, 1301 Center Ave, Fort Collins, CO 80523, USA; Abby.Ward@colostate.edu

2 Department of Chemical and Biological Engineering, Colorado State University, 1370 Campus Delivery, Fort Collins, CO 80523, USA; Sara.Dmytriw@rams.colostate.edu (S.D.); Ananya.Vajapayajula@rams.colostate.edu (A.V.)

* Correspondence: Christopher.Snow@colostate.edu

Abstract: Protein and DNA co-crystals are most commonly prepared to reveal structural and functional details of DNA-binding proteins when subjected to X-ray diffraction. However, biomolecular crystals are notoriously unstable in solution conditions other than their native growth solution. To achieve greater application utility beyond structural biology, biomolecular crystals should be made robust against harsh conditions. To overcome this challenge, we optimized chemical DNA ligation within a co-crystal. Co-crystals from two distinct DNA-binding proteins underwent DNA ligation with the carbodiimide crosslinking agent 1-ethyl-3-(3-dimethylaminopropyl)carbodiimide (EDC) under various optimization conditions: 5′ vs. 3′ terminal phosphate, EDC concentration, EDC incubation time, and repeated EDC dose. This crosslinking and DNA ligation route did not destroy crystal diffraction. In fact, the ligation of DNA across the DNA–DNA junctions was clearly revealed via X-ray diffraction structure determination. Furthermore, crystal macrostructure was fortified. Neither the loss of counterions in pure water, nor incubation in blood serum, nor incubation at low pH (2.0 or 4.5) led to apparent crystal degradation. These findings motivate the use of crosslinked biomolecular co-crystals for purposes beyond structural biology, including biomedical applications.

Keywords: co-crystal engineering; chemical ligation; bioconjugation; X-ray diffraction; DNA; DNA-binding protein

Citation: Ward, A.R.; Dmytriw, S.; Vajapayajula, A.; Snow, C.D. Stabilizing DNA–Protein Co-Crystals via Intra-Crystal Chemical Ligation of the DNA. *Crystals* **2022**, *12*, 49. https://doi.org/10.3390/cryst12010049

Academic Editor: Abel Moreno

Received: 5 December 2021
Accepted: 21 December 2021
Published: 30 December 2021

1. Introduction

Beyond serving as the fundamental components of life, proteins and DNA are also key building blocks for nanoscale self-assemblies. Biomolecular assemblies, ranging from 2D arrays to 3D crystals, are useful tools for structural biology, bio-catalysis, and biomedical applications [1–3]. Porous biomolecular crystals can even act as macromolecular scaffolds [4], providing structural details to guest macromolecules [5]. However, downstream applications of interest, including X-ray diffraction, are hindered by crystal fragility and intolerance to solvent conditions other than the crystal growth solution. In this study, we establish a protocol for the chemical ligation of DNA inside of crystals and we demonstrate structural resilience of crosslinked co-crystals which may further their application utility.

DNA assembly stability is a limiting factor for DNA nanotechnology and DNA crystals. While coding DNA sticky base overhangs can drive self-assembly, the non-covalent DNA base stacking interactions and Watson–Crick hydrogen bonds that stabilize the junctions are only stable under specific conditions. For example, crystallization conditions for DNA crystals typically feature high concentrations of divalent cations such as Mg(II) to balance the negative phosphate backbone of DNA [6]. DNA–protein co-crystals may be similarly reliant on counterions, particularly if counterions stabilize the DNA–protein binding event [7]. Crystal forms that bring DNA building blocks into close proximity are very sensitive to the counterion environment, and often dissolve or convert into a disordered aggregate when placed in water. To maximize application versatility, DNA structures should ideally be robust to solution variations, not just ionic strength but also temperature and pH [1]. Introducing covalent bonds across DNA–DNA interfaces has the potential to dramatically improve crystal macro-structure stability and could also improve X-ray diffraction.

Bioconjugation, or crosslinking, is a well-established strategy to improve the structural integrity of protein and DNA crystals [8]. The protein–protein interfaces found within protein crystals tend to be rich in primary amines and carboxylic acids. If all neighboring building blocks can be covalently linked, the resulting covalent organic framework can be a robust material. In traditional protein X-ray crystallography, glutaraldehyde, a highly reactive crosslinker, can increase crystal stability in varying solution conditions, and can even improve diffraction resolution [9,10]. In our previous work on protein crystals, we have found that glyoxal offers an effective alternative to glutaraldehyde [11,12]. Chemical crosslinking and photo-crosslinking methods for DNA crystals are also established in the literature [13–15]; however, we wanted to focus on a protocol in which the crosslinking does not require a specific sequence of DNA and does not add atoms to the structure (a zero-length crosslink).

Arguably the most natural form of sequence-independent DNA crosslinking is *ligation*, where the nicks dividing stacked dsDNA blocks are removed to generate longer contiguous DNA strands. For example, Li et al. used T4 DNA Ligase to ligate the DNA junctions within highly porous DNA crystals [16]. This elegant approach is limited to crystals that have large enough solvent channels for enzyme ingress. Here, we sought to optimize a chemical ligation alternative to the use of ligase that would be applicable to crystals with both large and small pores.

Our chemical ligation chemistry relies on 1-ethyl-3-(3-dimethylaminopropyl)carbodiimide (EDC), a water-soluble carbodiimide [8]. EDC is widely used, especially in protein conjugation, to crosslink primary amines to carboxylic acids. A less common chemistry for EDC is the activation of a terminal phosphate such that a suitably placed nucleophile can displace the leaving group [17]. When that nucleophile is the hydroxyl of a neighboring DNA strand, this chemistry results in a zero-length crosslink: a scar-less chemical ligation of DNA (Figure 1). EDC has been used to ligate dsDNA hairpins in solution [17], to link the phosphate backbone of stacked DNA in liquid crystals [18] and to stabilize a 600 nucleotide DNA origami structure [19]. Our work represents the first ligation via EDC of co-crystals containing protein and DNA. We show that EDC crosslinking dramatically increases crystal stability at the macroscale and does not prevent destroy the crystal nanostructure (i.e., treated crystals are still suitable for study via X-ray diffraction).

To demonstrate generality, we chemically ligate two different co-crystals of DNA-binding proteins containing stacked DNA–DNA interfaces (Figure 2). For convenience, we will refer to crystals of the RepE54 transcription factor bound to cognate 21-mer dsDNA as Co-Crystal One (CC1) (Figure 2A–D) and we will refer to crystals of the E2F8 transcription factor bound to cognate 15-mer dsDNA as Co-Crystal Two (CC2) (Figure 2E–G). The asymmetric unit for each co-crystal consists of a DNA-binding protein and short, cognate DNA duplex. Both co-crystals have existing models in the Protein Data Bank (PDB). CC1 is closely related to existing PDB entry 1rep, though the 1rep model corresponds to a crystal with differing DNA at the junction (Table S1). CC2 is identical to existing PDB entry 4yo2. The CC1 and CC2 crystals used in this study consist of dsDNA that is either blunt-ended or carries terminal 5′ or 3′ phosphates (Figure 3). In each co-crystal system, the crosslinking variables tested were terminal 5′ vs. 3′ phosphates, crosslinking time, EDC concentration, and repeated EDC dose. After EDC crosslinking, co-crystals had dramatically increased structural integrity with respect to changes in the solution condition.

To show foundational feasibility for biomedical applications, we demonstrated that crosslinked co-crystals remain robust in aqueous environments, blood serum, and at pH values found in the stomach (pH 2.0) or lysosomes (pH 4.5). Therefore, the EDC crosslinking results provided here may justify further investigation of chemically ligated co-crystals or pure DNA crystals as biomaterials. For scaffold-assisted crystallography [3] it is also important to note that the crosslinked co-crystals still diffracted X-rays. The crosslinked crystals tested here diffracted nearly as well as non-crosslinked crystals (anecdotally, a typical ~0.3 Å resolution difference). Additionally, we showed that this chemical ligation method is independent of the DNA sequence at the DNA–DNA junction. For example, despite differing DNA sequences at the junctions of CC1 and CC2, chemical ligation was effective in both cases. In summary, EDC ligation is a practical approach for crosslinking DNA inside of crystals and the optimized chemical crosslinking shown can provide the stability needed for diverse downstream applications.

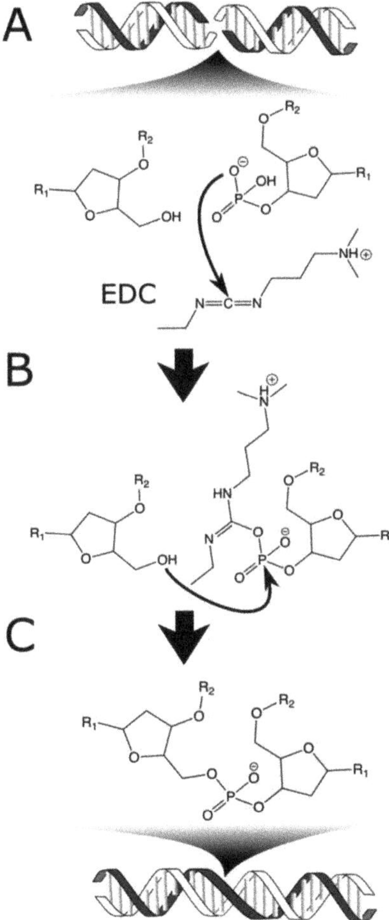

Figure 1. The mechanism of chemical DNA ligation with EDC. (**A**) A terminal 5′ hydroxyl and a terminal 3′ phosphate on neighboring DNA chains. The phosphate interacts with EDC to form an intermediate (**B**) and the hydroxyl displaces the reactive intermediate to form a zero-length crosslink (**C**) between the two DNA chains. R_1 is the nucleobase and R_2 is the phosphate backbone.

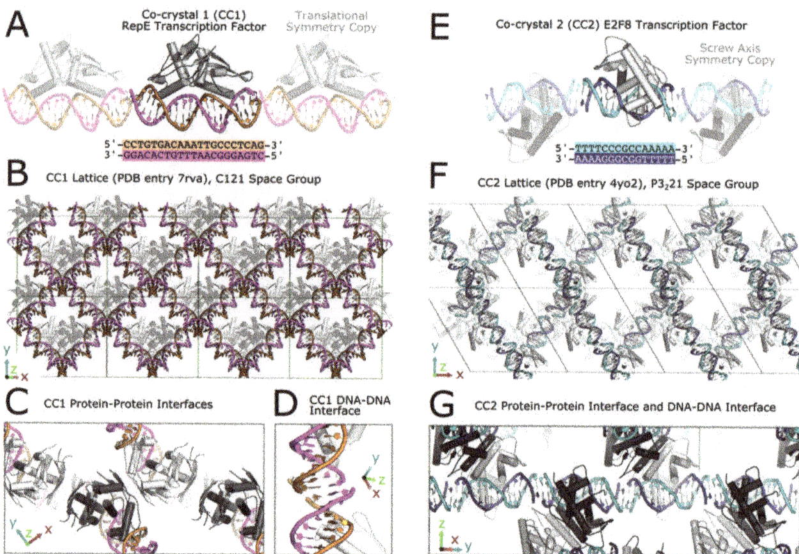

Figure 2. (**A**) The building block for co-crystal 1 (CC1) consists of the RepE54 transcription factor bound to 21-mer cognate DNA (represented here by PDB entry 1rep). (**B**) A collection of neighboring CC1 unit cells oriented to show the DNA stacks in 2 dimensions, with protein at 50% transparency. (**C**) The CC1 lattice has C121 symmetry, and all DNA–DNA junctions are symmetry equivalent to (**D**) the single DNA–DNA junction shown here. (**E**) The building block for co-crystal 2 (CC2) consists of the E2F8 transcription factor bound to 15-mer cognate DNA (represented here by PDB entry 4yo2). (**F**) A collection of neighboring CC2 unit cells oriented to show the DNA stacks in two dimensions, with protein at 50% transparency. (**G**) The CC2 lattice has $P3_221$ symmetry, and all DNA–DNA junctions are symmetry equivalent to the single DNA–DNA junction shown here. Images were generated in PyMOL.

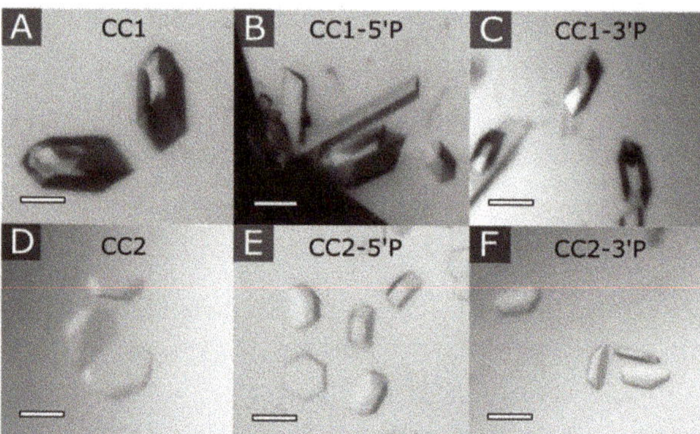

Figure 3. Examples of six co-crystal variants relative to a 100 micron scale bar. The CC1 crystals have C121 symmetry and tend to grow as monoclinic prisms: (**A**) CC1 without terminal phosphates, (**B**) CC1 with terminal 5′ phosphate, and (**C**) CC1 with terminal 3′ phosphate. In contrast, CC2 crystals have P3221 symmetry and tend to grow as truncated hexagonal prisms: (**D**) CC2 without terminal phosphates, (**E**) CC2 with terminal 5′ phosphate, and (**F**) CC2 with terminal 3′ phosphate.

2. Materials and Methods

2.1. Protein Cloning, Expression, and Purification

The protein sequence (Protocol S1) of RepE54 transcription factor (CC1 protein) from PDB code 1rep was cloned into a PSB3 vector with a N-terminal 6-Histag [20,21]. The Histone Source at Colorado State University expressed and purified CC1 protein as follows. *E. coli* CodonPlus RIPL competent cells were transformed with the CC1 protein expression plasmid and grown at 37 °C to a density of OD_{600} 0.6 in 2xYT broth containing Ampicillin (100 mg/L) and Chloramphenicol (25 mg/L). Isopropyl-β-D-thiogalactoside (IPTG) was added at 0.4 mM and the culture was continually shaken at 37 °C for 3 h. Cells were harvested by centrifugation and resuspended in PBS buffer supplemented with 300 mM NaCl, 0.2 mM AEBSF, and 5 mM B-mercaptoethanol and were homogenized by sonication at 50% output (10 cycles of 45 s on, 120 s off). Lysate was recovered by centrifugation at 27,000× *g* for 25 min. The supernatant was loaded onto Ni Excel Sepharose resins (CV = 15 mL, Cytiva), washed and eluted by a linear gradient of 0–500 mM imidazole in resuspension buffer. The fractions containing CC1 protein were pooled, concentrated using Amicon Ultra-15 10 kDa MWCO centrifugal filter unit (EMD Millipore) and loaded onto a size-exclusion HiLoad Superdex 200 PG column (Cytiva) equilibrated with sodium citrate buffer (100 mM Sodium citrate pH 6.2, 100 mM KCl, 10 mM $MgCl_2$ and 10% glycerol). Fractions containing CC1 protein were collected, concentrated to 15 mg/mL, and stored at −80 °C after freezing with liquid nitrogen.

The E2F8 transcription factor (CC2 protein) plasmid was graciously donated by the Taipale Lab (Protocol S1). The protein was expressed and purified based on previous guidelines [22]. CC2 protein with a TEV protease-cleavable N-terminal thioredoxin tag was expressed with a T7 promoter in *E. coli* BL21(DE3) cells. Upon addition of 0.5 mM IPTG, the cells were outgrown at 25 °C for 20 h. The cell pellets were sonicated in lysis buffer and applied to HisTrap (HisPur™ Ni-NTA Resin) equilibrated with HisTrap buffer (500 mM NaCl, 100 mM HEPES, 10 mM imidazole, 10% glycerol, 0.5 mM TCEP, pH 7.5). The protein was eluted with 200 mM imidazole in HisTrap buffer. CC2 protein was TEV cleaved from thioredoxin during dialysis using Snakeskin MWCO 10 kDa into HisTrap buffer. The cleaved product was separated from thioredoxin and TEV Protease by HisTrap, eluting with addition of HisTrap buffer. The CC2 protein was purified further with Nuvia™ cPrime™ Hydrophobic Cation Exchange Media, equilibrated with cation exchange buffer (50 mM NaCl, 100 mM HEPES, 10% glycerol, 0.5 mM TCEP, pH 7.5), and eluted with 100 mM NaCl in cation exchange buffer. The fractions containing CC2 protein were pooled, concentrated using Amicon Ultra-15 10 kDa MWCO centrifugal filter unit (EMD Millipore) and loaded onto a size-exclusion HiLoad Superdex 200 PG column (Cytiva) equilibrated with CC2 storage buffer (150 mM NaCl, 20 mM HEPES, 5% glycerol, 0.5 mM TCEP, pH 7.5). Size exclusion was completed at CSU's Histone Source. Fractions containing CC2 protein were collected, concentrated to 10 mg/mL, and stored at −80 °C after flash freezing with liquid nitrogen.

All protein sample purification was analyzed with SDS-PAGE (NuPAGE™ 4–12% Bis-Tris Gel) with MES SDS running buffer. Gels were stained with Imperial™ Protein stain. Protein concentrations were determined with Bradford Assay using Coomassie Plus™ Protein Assay Reagent.

2.2. DNA Duplex Annealing

DNA duplex sequences are given in Figure 2 and Table S1. The RepE54 co-crystal oligomers were designed from the original 22-mer in PDB code 1rep [20]. All sequences contained the 19 bp iteron sequence for DNA–protein binding, but the original duplex was truncated from a 22-mer to a 21-mer to eliminate an unresolved dangling base and to give a blunt ended DNA interaction for crosslinking. The E2F8 transcription factor co-crystal oligomers were the original duplex found in PDB code 4yo2 [22]. The CC1 and CC2 oligomers were synthesized and HPLC purified by Integrated DNA Technologies with termini containing no phosphates, 5′ phosphates, or 3′ phosphates. The oligomers were resuspended: CC1 oligomers in 50 mM Tris HCl, 100 mM KCl pH 7.0 and CC2 oligomers in 10 mM Tris base, 150 mM NaCl, 1 mM EDTA pH 7.5. The DNA duplexes were annealed by combining cognate ssDNA oligomers in a 1:1 molar ratio, heating to 94 °C for 2 min then slowly cooling to room temperature over approximately 60 min. The final concentration of CC1 and CC2 duplexes were 4 mM and 1 mM, respectively. DNA stocks were quantified with a Qubit4 (Qubit™ 1× dsDNA HS Assay Kit).

2.3. DNA–Protein Complex Co-Crystallization

All co-crystals were grown via sitting drop vapor diffusion. At 30 min prior to crystal plate setup, the protein and DNA were incubated at a 1:1.2 molar ratio. The DNA–protein complexes were kept on ice for 30 min prior to use. CC1, RepE54 transcription factor co-crystal, crystallization conditions were 30–120 mM $MgCl_2$, 2–16% PEG 400 and 100–220 mM Tris HCl pH 8.0. CC2, E2F8 transcription factor co-crystal, crystallization conditions were 40–300 mM ammonium sulfate, 5% PEG 400, 5–20% PEG 3350, and 80 mM HEPES pH 7.1. Crystals grew to a size of 50–150 μm^3 in a range of 24 h to 7 days.

2.4. EDC Crosslinking Co-Crystals

Co-crystals were washed in conditions similar to crystal growth conditions where growth buffer components that interfere with crosslinking were substituted (i.e., primary amines, carboxylic acids, and divalent cations). The CC1 wash solution consisted of 30–120 mM NaCl (substituting for $MgCl_2$), 2–16% PEG 400 and 100–220 mM MES pH 6.0 (substituting for Tris HCl pH 8.0). The CC2 wash solution consisted of 20–300 mM lithium sulfate (substituting for ammonium sulfate), 5% PEG 400, 10–30% PEG 3350 (an increase of 10% PEG 3350 compared to the growth solution), and 80 mM MES pH 6.0 (substituting for HEPES pH 7.1). The 10% additional PEG 3350 for CC2 appeared to prevent the crystals from degrading upon addition of the wash. The co-crystals were washed in 9-well glass plates (Hampton) to remove additional protein and DNA monomers and unwanted buffer components. 1-Ethyl-3-(3-dimethylaminopropyl)carbodiimide (EDC) (Advanced Chemtech CAS#:25952-53-8) was resuspended in the wash solution to final concentration values ranging from 5 to 80 mg/mL and used immediately. The co-crystals were crosslinked in a 200 µL EDC solution volume for varying time points. The co-crystal crosslinking reaction was quenched by moving crystals to 1× Tris-Borate-EDTA (TBE) buffer pH 8.3 containing 3.5 M urea.

2.5. DNA Gel Electrophoresis and Densitometry

Crosslinked co-crystals were dissolved in 3.5M Urea in 1× TBE supplemented with Proteinase K and incubated at 50 °C overnight. When crystals were too robust to dissolve under these harsh conditions, the crystals were heated to 94 °C for 1 h and glass crystal crushers (Hampton) were used to crush the crystals prior to chemical and enzymatic attack. The crystals were analyzed with 10% or 15% Novex™ TBE-Urea Polyacrylamide Gel Electrophoresis (PAGE) with 1× TBE running buffer. DNA ladders were GeneRuler Low Range DNA Ladder (Thermo Scientific, Houston, USA) for CC1 gels and Ultra Low Range DNA Ladder (Invitrogen) for CC2 gels. The control lanes included 1-mer dsDNA and 2-mer dsDNA, prepared by annealing oligos as mentioned in Section 2.3 (Table S1 Duplex IDs 1.1, 1.4, 2.1, and 2.4). Gels were incubated with 3 × GelRed™ Nucleic Acid Gel Stain and imaged with a UVP Bioimaging System on the Ethidium Bromide setting. For further validation, selected ligation products for CC1 were also analyzed with a TapeStation D1000 ScreenTape assay (Agilent) (Figure S1) at CSU's Next Generation Sequencing Core. The gels and TapeStation were analyzed via densitometry.

2.6. DNA Gels and Densitometry

For densitometry, we used ImageJ (1.52 k) to obtain raw x,y,intensity values for the gels shown in Section 3.1. We averaged these data over x values and used custom Python scripts (within "cocrystal_ligation_scripts.zip" hosted on Zenodo [23]) as well as the lmfit module [24] to obtain non-linear best fits of the gel intensity. Specifically, we modeled peaks using Gaussian functions. We also modeled the background using diffuse Gaussian functions. Crystals with more crosslinking produced overlapping gel bands for higher-order ligation products. One of the benefits of using a mathematical curve fitting framework is our ability to fit (albeit approximately) these populations. Specifically, we fit the peak position trend using the well-separated gel bands corresponding to smaller ligation products. Then, we fit the highly overlapping region using extrapolated peak positions with fitting parameter restrictions implemented via lmfit. Inspection of the fitting results (Figure S2) gave us confidence that higher-order band intensity fit was reasonable.

In principle, longer DNA ligation products can adsorb a greater number of GelRed fluorophores, proportionally with the DNA length. Ignoring this effect might cause us to overestimate the ligation yield. Accordingly, we proceeded to normalize the estimated molar ratio of the ligation products (Section 3.2) by dividing each band intensity by the assigned DNA block size (divide by N for N-mer DNA blocks). The raw band intensity fits are provided in Table S2.

2.7. Random Ligation Model

As exemplified in Section 3.2, the densitometry data could be interpreted in terms of the relative population of unfused DNA, ligated 2-mer, ligated 3-mer, etc. We sought to interpret these data in terms of the likely percentage of the dsDNA-dsDNA interfaces that have gained at least one covalent bond via EDC ligation. First, we used the estimated molar ratio of products from gel densitometry to estimate the fraction of potential ligation sites that were ligated. Second, to compute the expected distribution of fused DNA blocks of varying length, we implemented a simple 1D simulation in Python (Protocol S2, also within "cocrystal_ligation_scripts.zip" [23]) in which all 85712 nicks between DNA blocks in a 1D stack of 42,857 blocks (a 300 micron stack) were equally likely to be randomly removed in each unit of time. This "random ligation model" is arguably the least complex theoretical model for the crosslinking process, ignoring transport phenomena and assuming that all possible ligation sites throughout the crystal undergo ligation randomly with equal probability per unit of time. We also developed a biased ligation model in which sites near the crystal interior are less likely to be ligated than sites near the crystal surface (Protocol S3, also within "cocrystal_ligation_scripts.zip" [23]).

2.8. X-ray Diffraction Data Collection, Refinement and Omit Maps

Single-crystal X-ray diffraction (XRD) data were collected for CC1 crystals containing 5′ and 3′ terminal phosphates. Crosslinked crystals with 3′ terminal phosphates were also analyzed via XRD. Crystals were briefly swished through cryo-protectant solution (300 mM $MgCl_2$, 30% PEG 400, and 100 mM Tris HCl pH 8.0) and flash-frozen in liquid nitrogen. Frozen crystals were stored in Rigaku ACTOR Magazines (Mitegen) and shipped to the Advance Light Source Beamline 4.2.2 for data collection. Full datasets were collected on a CMOS detector from 0 to 180 degrees with an omega delta of 0.2° and an exposure time of 0.3 s. Data were processed with XDS [25] and molecular replacement and refinement within PHENIX [26] and COOT [27]. As a result, the original co-crystal for RepE54 transcription factor (2.60 Å PDB code 1rep) was updated with a higher-resolution structure (1.89 Å PDB code 7rva). The updated structure was solved with molecular replacement using the PDB code 1rep. CC1 crystal structures containing 5′ or 3′ terminal phosphates were solved via molecular replacement in PHENIX using the updated original CC1 as a starter model. For all structures, the same R-free flags were used during refinement in PHENIX and COOT. Structure factor data were truncated using I/sigma(I) >1.5 as a cutoff. The resulting structures were of: CC1 with terminal 5′ phosphates (PDB code: 7sgc), CC1 with terminal 3′ phosphates (PDB code: 7sdp), low EDC crosslinked (5 mg/mL EDC, 12 h) CC1 with terminal 3′ phosphates (PDB code: 7soz), and heavy EDC crosslinked (30 mg/mL EDC, 12 h, two doses) CC1 with terminal 3′ phosphates (PDB code: 7spm). Standard X-ray diffraction data quality statistics are provided in Tables 1 and 2.

Omit maps were generated for each structure, shown in Section 3.3. To prevent bias of the electron density at the junctions in crosslinked structures, discovery and omit maps were generated with structures containing no terminal phosphates. After generating discovery and omit maps, the terminal phosphates were added to the structures and refined for submission to the PDB. In the final PHENIX refine of heavy crosslinked CC1 terminal 3′ phosphates, a custom geometry bond restraint was added because the electron density indicated ligation at both junctions. The terminal 3′P and flanking 5′OH were given a bond length restraint of 1.59 Å, the ideal length of the phosphate-oxygen bond in the DNA backbone [28].

Table 1. X-ray diffraction statistics for the updated original CC1 crystal, the CC1 crystal with terminal 5′ phosphates, and the CC1 crystal with terminal 3′ phosphates.

	Updated CC1 Original PDB Code 7rva	**CC1 5′p PDB Code 7sdp**	**CC1 3′p PDB Code 7sgc**
Data collection			
Light source	Synchrotron	Synchrotron	Synchrotron
Wavelength (Å)	1.0	1.0	1.0
Resolution range (Å)	33.78–1.89 (1.958–1.89) *	34.45–2.7 (2.796–2.7) *	37.68–3.01 (3.118–3.01) *
Space group	C 1 2 1	C 1 2 1	C 1 2 1
Unit cell dimensions			
a, b, c (Å)	107.578 80.715 73.299	109.161 83.051 74.349	109.142 83.089 74.593
α, β, γ (°)	90 122.63 90	90 123.948 90	90 122.691 90
Total reflections	152,923 (15,187)	51,877 (5051)	40,979 (3834)
Unique reflections	41,778 (2501)	14,436 (1425)	11,132 (1074)
Multiplicity	3.7 (3.6)	3.6 (3.5)	3.7 (3.5)
Completeness (%)	90.98 (59.52)	93.97 (92.88)	98.37 (94.79)
Mean I/sigma(I)	10.77 (1.71)	13.56 (1.69)	11.45 (1.67)
Wilson B-factor	32.28	61.1	57
R-merge	0.05735 (0.7813)	0.06375 (0.7563)	0.1406 (1.383)
R-meas	0.06726 (0.9191)	0.07496 (0.8954)	0.1652 (1.638)
R-pim	0.03484 (0.4793)	0.03905 (0.4741)	0.08579 (0.8663)
CC1/2	0.998 (0.749)	0.997 (0.669)	0.993 (0.574)
CC *	1 (0.926)	0.999 (0.895)	0.998 (0.854)
Refinement			
Reflections used in refinement	38,499 (2501)	14,342 (1422)	11,060 (1074)
Reflections used for R-free	2007 (132)	752 (78)	586 (55)
R-work	0.2074 (0.5745)	0.1866 (0.3206)	0.1820 (0.3258)
R-free	0.2490 (0.5833)	0.2457 (0.4038)	0.2351 (0.3987)
CC (work)	0.972 (0.374)	0.968 (0.792)	0.973 (0.730)
CC (free)	0.937 (0.202)	0.964 (0.578)	0.929 (0.546)
Number of non-hydrogen atoms	2970	2800	2825
Macromolecules	2766	2770	2776
Ligands	3	3	2
Solvent	201	27	47
Protein residues	228	230	232
RMS (bonds) (Å)	0.009	0.011	0.012
RMS (angles) (°)	1.11	1.3	1.35
Ramachandran favored (%)	98.14	95.85	95.43
Ramachandran allowed (%)	1.86	4.15	4.57
Ramachandran outliers (%)	0	0	0
Rotamer outliers (%)	0	0	0
Clashscore	3.14	8.45	8.41
Average B-factor	55.73	75.32	62.28
Macromolecules	56.07	75.49	62.69
Ligands	55.14	78.12	52.05
Solvent	51.08	57.85	38.7
Number of TLS groups	10	10	10

* Values in parentheses are for high-resolution shell.

Table 2. X-ray diffraction statistics for the CC1 crystal with terminal 3′ phosphates and low crosslink (5 mg/mL EDC for 12 h) and the CC1 crystal with terminal 3′ phosphates and heavy crosslink (two doses of 30 mg/mL EDC for 12 h).

	CC1 3′p Low EDC Crosslink 5 mg/mL EDC for 12 h PDB Code 7soz	**CC1 3′p Heavy EDC Crosslink 2 Doses of 30 mg/mL EDC for 12 h PDB Code 7spm**
Data collection		
Light source	Synchrotron	Synchrotron
Wavelength (Å)	1.0	1.0
Resolution range (Å)	37.05–3.14 (3.252–3.14) *	33.78–3.28 (3.397–3.28) *
Space group	C 1 2 1	C 1 2 1
Unit cell dimensions		
a, b, c (Å)	111.993 79.181 74.903	110.817 80.174 74.723
α, β, γ (°)	90 123.023 90	90 122.899 90
Total reflections	34,361 (3280)	31,305 (3106)
Unique reflections	9433 (884)	8467 (830)
Multiplicity	3.6 (3.6)	3.7 (3.7)
Completeness (%)	95.82 (90.02)	97.95 (93.58)
Mean I/sigma(I)	10.37 (2.22)	8.42 (2.09)
Wilson B-factor	83.36	94.48
R-merge	0.08469 (0.5434)	0.1187 (0.6541)
R-meas	0.09971 (0.6391)	0.1394 (0.7672)
R-pim	0.05208 (0.3334)	0.0724 (0.3973)
CC1/2	0.997 (0.946)	0.998 (0.94)
CC *	0.999 (0.986)	0.999 (0.984)
Refinement		
Reflections used in refinement	9290 (857)	8356 (802)
Reflections used for R-free	498 (49)	448 (40)
R-work	0.1936 (0.3019)	0.2010 (0.3183)
R-free	0.2637 (0.3575)	0.2528 (0.3556)
CC (work)	0.982 (0.891)	0.985 (0.869)
CC (free)	0.915 (0.899)	0.987 (0.932)
Number of non-hydrogen atoms	2691	2649
Macromolecules	2687	2641
Ligands	1	1
Solvent	3	7
Protein residues	224	221
RMS (bonds) (Å)	0.012	0.015
RMS (angles) (°)	1.49	1.63
Ramachandran favored (%)	91.28	91.16
Ramachandran allowed (%)	8.72	8.37
Ramachandran outliers (%)	0	0.47
Rotamer outliers (%)	0	0
Clashscore	14.36	14.92
Average B-factor	109.8	130.22
Macromolecules	109.91	130.43
Ligands	30	38.22
Solvent	38.28	63.68
Number of TLS groups	10	10

* Values in parentheses are for high-resolution shell.

2.9. Stability Assays

Crystals were crosslinked using the Section 2.4 protocol, with 15 mg/mL EDC for 20 h. The EDC reaction was quenched in 50 mM Tris base pH 8.0 for 30 min. The crystals were equilibrated in crosslinking wash solution for 30 min prior to looping to stringent conditions. The stability test buffers used were as follows: molecular biology grade water (CORNING), very low pH 2.0 0.01 M HCl buffer (to mimic stomach acid), a moderately low pH 4.5 citrate buffer (46 mM sodium citrate, 54.1 mM citric acid to mimic lysosomal fluid pH), and blood serum (HyClone, bovine calf serum). Pictures for each trial are in Figures S3–S6. Crystal pictures were obtained with a Moticam 3.0 MP camera attached to a Motic SMZ-168 stereozoom microscope and crystal measurements were performed in Motic Images Plus 2.0 (Figure S4 and Protocol S4).

3. Results

3.1. Chemical Ligation in Co-Crystals

Within our two co-crystal families (CC1 and CC2), we observed clear evidence of chemical ligation of stacked DNA duplexes. Both co-crystals demonstrated broadly similar ligation results, emphasizing the generality of this ligation method to co-crystals in which blunt-ended DNA blocks are suitably positioned to resemble contiguous DNA. As shown in Table 3, the PDB entries for the parent structures of both CC1 (7rva) and CC2 (4yo2) have junction step geometry that is reasonably comparable to contiguous B-DNA as calculated using x3DNA [29]. Except for the twist and roll across the CC2 junction (as seen in PDB entry 4yo2), all step geometry parameters are within 2 standard deviations of the B-DNA mean. It is possible that other co-crystals in which the DNA–DNA junctions have a geometry less like contiguous DNA would resist ligation. Additionally, in the preliminary crosslinking tests shown here, the crosslinking was successfully independent of the sequence at the DNA ends. CC1 has GC/CG flanking ends while CC2 has AT/TA flanking ends. The sequence independence of this ligation strategy is advantageous for DNA structure design projects where the junction sequence may be constrained for functional reasons. Table 3 also reports an interesting asymmetry between the two nick sites at the DNA–DNA junctions within the CC1 family of structures. We report the distance between C5′ and O3′ to avoid relying on the less certain O5′ position. For calibration, an idealized B-DNA model from x3dna had C5′ to O3′ distances of 2.73 Å for contiguous bases, but this span is variable (2.99 ± 0.17 Å) elsewhere within the dsDNA of PDB entry 7rva. One of the two CC1 nick sites, chain B, was invariably closer than chain A (e.g., 3.75 Å rather than 4.22 Å in CC1-3′P), and electron density suggested that this shorter gap (chain B) was more readily ligated.

Table 3. DNA–DNA junction geometry parameters before and after phosphorylation and ligation. The likelihood of successful chemical ligation for stacked DNA may depend on geometry details across the junction. Here, we compare the geometry of the junction in the parent PDB models for CC1 and CC2, as well as the blunt-ended 5′ or 3′ phosphorylated CC1 crystals, to the geometry of contiguous bases in idealized B-DNA from Olson et al., 1998 [30]. The junctions are not symmetric, and differing distances for the two nicks across the junctions are also shown.

Junction Parameters from x3dna	CC1	CC2	B-DNA *	CC1-5′P	CC1-3′P	CC1-3′P EDC Heavy
PDB code	7rva	4yo2		7sgc	7sdp	7spm
Base pair step parameters	GC/CG	AT/TA		GC/CG	GC/CG	GC/CG
Shift (Å)	−0.03	0.36	0.0 ± 0.51	0.10	0.05	0.50
Slide (Å)	−0.81	−1.27	0.35 ± 0.78	−1.00	−2.03 †	−0.52
Rise (Å)	3.49	3.61	3.32 ± 0.19	3.78 †	3.56	4.02 †
Tilt (°)	1.87	0.91	0.0 ± 3.4	2.51	0.72	1.72
Roll (°)	1.25	−15.84 †	1.4 ± 5.1	1.94	4.03	2.10
Twist (°)	36.64	20.95 †	35.4 ± 6.3	39.10	28.18	36.67
Nick distances C5′ to O3′						
Chain A: (C for CC2)	3.64	3.71	2.73	4.01	4.22	3.46 ‡
Chain B: (D for CC2)	3.45	3.86	2.73	3.39	3.75	3.30 ‡

* Base pair step parameters from Olson et al. 1998 [30]. C5′ to O3′ distance from x3dna idealized B-DNA. † Values differ from B-DNA by more than 2 standard deviations. ‡ Values are the distances in the refined "discovery" models prior to addition of the 3′ phosphate (not PDB 7spm).

EDC crosslinking was tested for both 5′ and 3′ phosphate laden crystals. For CC1, the 3′ phosphate resulted in superior ligation yield than the 5′ phosphate in each trial (Figures 4 and S7–S11). On the other hand, CC2 ligation yields had a modest difference in the ligation yield for 3′ and 5′ phosphates. Given the limited dataset, it is premature to conclude that 3′ phosphates will typically give a higher ligation yield within co-crystals.

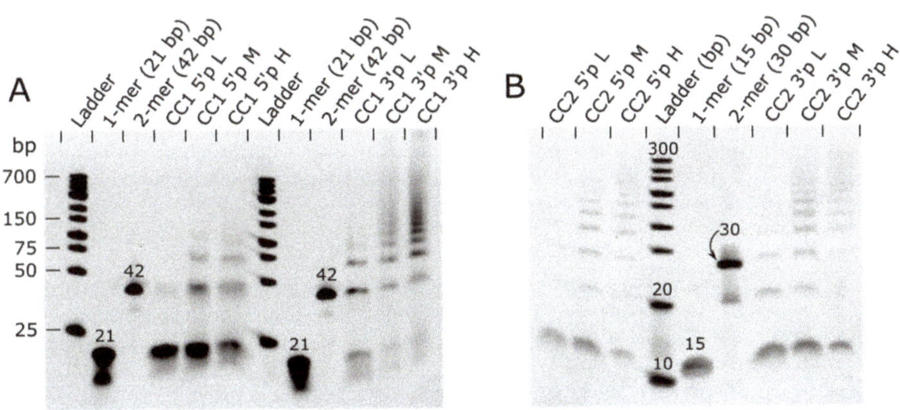

L = Low Crosslink (1 dose 5 mg/mL EDC 12h)

M = Moderate Crosslink (1 dose 30 mg/mL EDC 12h)

H = Heavy Crosslink (2 doses 30 mg/mL EDC 12h)

Figure 4. TBE-urea gels of (**A**) CC1 and (**B**) CC2 chemical ligation. In both co-crystals, additional ligation was achieved with increased EDC concentration and a second EDC dose. (**A**) A 10% TBE-urea gel of CC1 illustrating a much-improved ligation product distribution for 3′ vs. 5′ phosphates. (**B**) A 15% TBE-urea gel of CC2 illustrating a modestly improved ligation product distribution for 3′ vs. 5′ phosphates. Assigned band sizes are given in bp.

In both systems, DNA ligation was dependent on the presence of the terminal phosphates as well as on the crystal template; control crystals lacking terminal phosphates yielded no observable ligation products (Figure S10). Additionally, freely diffusing DNA blocks carrying terminal phosphates (but lacking the co-crystal scaffold) also yielded no observable ligation products when exposed to EDC (Figure S10). This second control demonstrated that the scaffold was necessary for ensuring efficient ligation of blunt-ended DNA blocks. The absence of observable ligation for building blocks in the absence of the crystal "scaffold", precludes a systematic study of the effects of precursor ligation on crystal growth. Future work will determine, as a function of sticky overhang length, the extent to which blocks with sticky overhangs can be ligated within crystals and in solution.

Crosslinking reaction time was clearly and directly related to ligation reaction yield during the first 12 h (Figure S7). It was less clear if reaction yield was further improved by incubation beyond 12 h. Therefore, 12 h crosslinking incubations were used for the subsequent ligation optimization trials.

In the next series of experiments, we optimized EDC concentration for maximum ligation yield. We assayed the ligation product distribution as a function of concentration from 5 mg/mL EDC to 80 mg/mL EDC. As hypothesized, increasing the concentration of EDC increases the ligation of DNA duplexes in the co-crystals (Figures 4 and S8). In CC1 trials, we did not see a noticeable increase in ligation beyond 30 mg/mL. However, in CC2 trials, there was improved ligation at 60 mg/mL. We also subjected the co-crystals to multiple fresh doses of EDC (30 mg/mL) to determine if we could achieve near 100% ligation. For both co-crystal systems, multiple doses of EDC did increase ligation yields (Figure S9) but did not approach 100% ligation yields.

Reaction buffer components were critical for successful ligation. We observed, at the outset of this project, that the presence of magnesium chloride in the crosslinking buffer appeared to interfere with the crosslinking reaction. This was problematic because the CC1 crystal growth conditions contain a significant amount of magnesium chloride. In our crystallization trials, 30–120 mM magnesium chloride was required for growth [21]. Additionally, there is a structural Mg(II) at the DNA–protein interface coordinated by Glu77 and Asp81. To circumvent the apparent deleterious role of Mg(II) on CC1 crosslinking, we replaced magnesium chloride with sodium chloride in the wash solution for all CC1 crosslinking trials. At the conclusion of the project, we again confirmed that Mg(II) was deleterious to ligation by adding Mg(II) to the optimized ligation protocol. Specifically, we verified that supplementing the crosslinking incubation buffer with 90 mM or 110 mM $MgCl_2$ noticeably reduced the ligation yield (Figure S11). The exact role of Mg(II) in inhibiting the ligation reaction is not clear, but might involve reduced availability of the nucleophilic phosphate groups.

3.2. Ligation Model Compared to Experimental Co-Crystal Ligation

The ligation product distributions we experimentally obtained should shed light on the stochastic process of ligation. Using densitometric analysis of electrophoresis results, we quantified the population ratio of bands assigned to non-modified DNA blocks as well as fused 2-mer, 3-mer, etc. For selected gels, we also obtained TapeStation results (Figure S1). The relative population of the end-product distribution was fairly consistent for gel band populations measured with TBE-urea gels in ImageJ compared to the automated TapeStation analysis (Figure S1).

Next, we sought to calculate a global performance metric for the ligation yield, P_{LIG}, as the fraction of all possible DNA–DNA nick sites throughout a crystal that were ligated. One destructive assay to quantify the ligation yield throughout an entire crystal is to analyze the implications of the final DNA product distribution recovered after the crystal is dissolved and the protein components are removed. A related quantity is P_{DSB}, the probability that any random DNA–DNA junction within the crystal remains a double-strand break (DSB). If we count the number of DNA blocks of each length (n_i) present in the crystal, we ignore edge effects and estimate the total number of DSB as $N_{DSB} = \sum_i n_i$. For the same crystal, the estimated total number of original junctions (regardless of final ligation status) would be $N_{JXN} = \sum_i i{\cdot}n_i$. For example, adding a single fused 3-mer to the crystal increases the DSB tally by one, but increases the tally of all possible junctions by three. Then, to compute the total probability of encountering DSB, we calculate:

$$P_{DSB} = \frac{N_{DSB}}{N_{JXN}} = \frac{\sum_i n_i}{\sum_i i{\cdot}n_i} = \frac{\sum_i n_i / \sum_i n_i}{\sum_i i{\cdot}n_i / \sum_i n_i} = \frac{1}{\sum_i i{\cdot}x_i} \quad (1)$$

In the final equation, x_i is the mole fraction for the DNA block of length i. Therefore, to estimate the P_{DSB}, we can use estimated mole fractions from electrophoresis and densitometry (Figures 4 and S2 and Table 4). Accurately calculating P_{DSB} does require including the small mole fractions for higher-order products (Table S3) since longer products contribute proportionally more to $\sum_i i{\cdot}x_i$. To estimate the uncertainty in each P_{DSB}, we used 500 numerical trials in which random noise was added to $i{\cdot}x_i$ to mimic densitometry measurement error. We used noise comparable to $i{\cdot}x_i$ for the highest-order ligation products (normal variate with standard deviation 0.03), such that the smallest $i{\cdot}x_i$ values would regularly fall to 0 after the addition of random noise.

While the probability of encountering a double-strand break in the crystal (P_{DSB}) is an important parameter, it would also be useful to know P_{LIG}, the probability of each terminal phosphate having undergone ligation. In the context of the random ligation model (RLM), ligation events throughout the crystal are independent and occur with equal probability at all nick sites. Therefore, the incidence of double-strand breaks within the crystal should occur with the joint probability of independent events, $P_{DSB} = (1 - P_{LIG})^2$. Thus, the overall probability that a random terminal phosphate within the crystal will be ligated is $P_{LIG} = 1 - \sqrt{P_{DSB}}$. The joint probability that DNA junctions will be double ligated is $P_{DLIG} = 1 - 2\sqrt{P_{DSB}} + P_{DSB}$, and the probability that they will be singly ligated is $P_{SLIG} = 2(\sqrt{P_{DSB}} - P_{DSB})$.

Table 4. Distribution of DNA block sizes as a function of crosslinking protocol and 3′ vs. 5′ terminal phosphates. The data shown correspond with the gel lanes in Figure 4. The crosslinking protocols low, medium, and high were 1 dose of 5 mg/mL EDC for 12 h, 1 dose of 30 mg/mL EDC for 12 h, and 2 doses of 30 mg/mL EDC for 12 h each, respectively. The values in this table are weighted so that the DNA length and dye intensity contributes to the final value. Unweighted values are found in Table S2. The full table including estimated mole fractions for higher-order products is found in Table S3. PDSB, PSB, and PLIG were calculated for each crosslinked crystal sample. Uncertainties are standard deviations in derived quantities after 500 trials in which noise (standard deviation 0.03) is introduced into relative band intensities.

Parent Crystal	**CC1-3′P**	**CC1-3′P**	**CC1-3′P**	**CC1-5′P**	**CC1-5′P**	**CC1-5′P**
Crosslinking Protocol	low	medium	high	low	medium	high
DNA block size	[%]	[%]	[%]	[%]	[%]	[%]
1	58.7	30.0	24.9	98.6	91.6	82.1
2	18.7	16.8	14.9	1.4	7.3	9.9
3	15.2	15.3	15.6		1.0	6.3
4	5.0	11.0	11.0		0.2	1.5
5	2.4	6.3	8.4			0.2
6		6.5	6.9			
7		4.4	5.8			
8 and above		9.7	12.5			
P_{DSB} *	0.58 ± 0.01	0.27 ± 0.01	0.24 ± 0.01	0.98 ± 0.01	0.90 ± 0.02	0.78 ± 0.02
$P_{SLIG} = 2(\sqrt{P_{DSB}} - P_{DSB})$.	0.37 ± 0.01	0.50 ± 0.00	0.50 ± 0.00	0.02 ± 0.01	0.09 ± 0.02	0.21 ± 0.02
$P_{DLIG} = 1 - 2\sqrt{P_{DSB}} + P_{DSB}$	0.06 ± 0.00	0.23 ± 0.01	0.26 ± 0.01	0.00 ± 0.00	0.00 ± 0.00	0.01 ± 0.00
$P_{LIG} = 1 - \sqrt{P_{DSB}}$	0.24 ± 0.01	0.48 ± 0.01	0.51 ± 0.01	0.01 ± 0.01	0.05 ± 0.01	0.12 ± 0.01
Parent Crystal	**CC2-3′P**	**CC2-3′P**	**CC2-3′P**	**CC2-5′P**	**CC2-5′P**	**CC2-5′P**
Crosslinking Protocol	low	medium	high	low	medium	high
DNA block size	[%]	[%]	[%]	[%]	[%]	[%]
1	94.4	80.3	74.4	96.9	84.8	72.2
2	2.6	4.8	3.3	1.2	5.6	3.1
3	1.5	4.5	4.4	1.9	4.9	7.7
4	0.8	3.6	3.7		2.6	5.2
5	0.7	2.5	2.9		1.1	2.9
6		1.3	2.9		1.0	2.7
7		1.3	2.3			1.9
8 and above		1.7	6.1			4.3
P_{DSB} *	0.90 ± 0.03	0.60 ± 0.03	0.45 ± 0.02	0.95 ± 0.02	0.75 ± 0.02	0.48 ± 0.01
$P_{SLIG} = 2(\sqrt{P_{DSB}} - P_{DSB})$.	0.10 ± 0.03	0.35 ± 0.02	0.44 ± 0.01	0.05 ± 0.02	0.23 ± 0.02	0.43 ± 0.01
$P_{DLIG} = 1 - 2\sqrt{P_{DSB}} + P_{DSB}$	0.00 ± 0.00	0.05 ± 0.01	0.11 ± 0.01	0.00 ± 0.00	0.02 ± 0.00	0.10 ± 0.01
$P_{LIG} = 1 - \sqrt{P_{DSB}}$	0.05 ± 0.02	0.22 ± 0.02	0.33 ± 0.01	0.02 ± 0.01	0.13 ± 0.01	0.31 ± 0.01

* Calculated from experimental mole fractions per Equation (1). Other probabilities are calculated as shown.

This analysis of the electrophoresis experiments suggests that ~50% of the terminal phosphates within the most thoroughly crosslinked CC1-3′P crystal have undergone ligation. On the other hand, ~75% of the DNA–DNA junctions in this crystal had at least one ligated chain. In summary, the ligation product ratio analysis suggests that a moderate fraction of the phosphates within these crystals have undergone the target ligation reaction, leading to an important question. What factors are limiting the yield? Incomplete ligation could result if a random population of terminal phosphates are missing, or otherwise incapable of on-target ligation. We used simulations to verify that the predicted RLM product ratio did not change when we postulated that a random subset of nick sites is incapable of ligation. This makes sense because junctions that are randomly selected to be incapable of ligation are functionally equivalent to sites that are randomly selected to be ligated last (i.e., after we stop ligating since we have reached PDSB).

It may also be possible that ligating one phosphate at a DNA–DNA junction would negatively affect neighboring ligation probabilities. However, evidence for such allostery is lacking. Instead, the observed product distributions for CC1 ligation outcomes (Table 4), were close to the distributions predicted by the RLM (Figure S12). One small but consistent deviation from the RLM was a lower 2-mer, and higher 3-mer population than predicted. This observation seems to preclude the simplest negative allostery scenario (where one ligation event would reduce the probability at flanking sites). We cannot rule out the possibility that this discrepancy is an artifact associated with the gel electrophoresis densitometry.

The CC2 ligation outcomes (Table 4) were significantly less consistent with distributions predicted by the RLM. Once more, the 3-mer population was often higher than expected, frequently exceeding the 2-mer population (which never happens in the RLM). This effect also seemed to extend to anomalously common 4-mers. A more striking divergence from the RLM prediction was the high population of non-ligated 1-mer blocks. Regardless of the RLM fit, the significant difference between the 1-mer mole fractions and the PDSB values obtained from all the mole fractions strongly implicates that the RLM is lacking.

To investigate, we tested biased ligation model simulations. One possible explanation is that the ligation outcomes were driven partially by kinetics and molecular transport phenomena. Hypothetically, ligation sites near the crystal exterior might be more likely to be ligated than possible sites near the crystal center since reactive molecules must traverse the outer layers to react the interior. To determine the likely implications of this scenario, we conducted biased random ligation simulations (Protocol S3) that increased the probability of ligation events near the surface, decreased the probability at the center, and terminated the random ligation process at a set PDSB threshold. Perhaps counterintuitively, this spatial bias increased the predicted 1-mer mole fraction. A high 1-mer fraction is partially consistent with the observed product distribution for CC2. The overall lower ligation yield achieved for CC2 crystals compared to CC1 is also consistent with the hypothesis that the CC2 crystal interior is systematically under-ligated.

3.3. *Ligation Structural Details*

Co-crystal structural details were revealed with X-ray diffraction at the Advanced Light Source beamline 4.2.2. Electrophoresis data (Figure 4) suggest that the CC2 DNA is stacked as intended. However, while high-resolution diffraction for CC2 crystals should be possible (3.07 Å reported by Morgunova et al. (22)), our CC2 crystals have, to date, yielded poor diffraction (>10 Å). Therefore, we chose to focus on the CC1 crystals as the model crystals to observe ligation via X-ray diffraction".

Here, we report five new crystal structures for CC1. We obtained a 1.89 Å dataset for the original co-crystal, which revealed additional details beyond the original model (PDB code: 1rep, 2.60 Å). Komori et al. varied the DNA building block to optimize resolution (20), finding that dangling Ts resulted in the best data. Our updated structure provides a rationale for this empirical observation. Specifically, one of the dangling T bases is resolved, and participates in a crystallographic contact. Removing the dangling Ts decreased the resolution of our native structures from 1.9 Å to 2.7 Å (CC1-5′p) or 3.01 Å (CC1-3′p). Once crystals were crosslinked with low (15 mg/mL 12 h) and heavy (2 doses 30 mg/mL 12 h) EDC, the crystals maintained diffraction, albeit with a moderate loss in diffraction (3.14 Å and 3.28 Å, respectively).

Models were refined with PHENIX [26] and COOT [27]. The electron density for the heavily ligated DNA junction was consistent with contiguous DNA, despite omitting the terminal phosphate throughout prior refinement calculations. Figure 5 shows omit maps where any terminal phosphates are omitted, along with the bases flanking the junctions. The potential for overlapping electron density contributions from non-ligated and ligated phosphates makes it difficult to quantify occupancy. Nonetheless, we observed clear trends. Prior to ligation, the positions of 3′ phosphates (Figure 5C) or 5′ phosphates (Figure 5D) were reasonably clear.

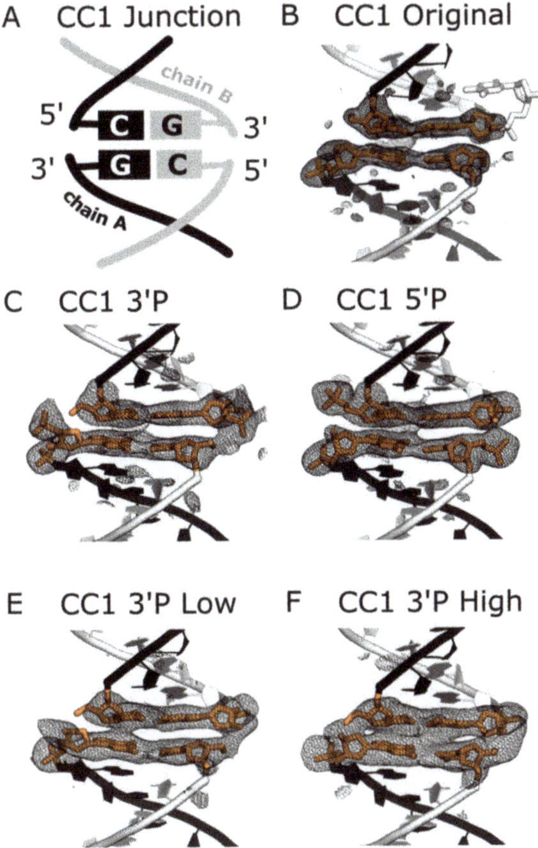

Figure 5. Omit maps for (**A**) the CC1 DNA–DNA junction. (**B**) Our updated model for the original structure resolves one of the dangling 5′ terminal bases (white sticks). Prior to ligation, CC1 crystals grow with either (**C**) terminal 3′ phosphates or (**D**) terminal 5′ phosphates. Whereas (**E**) low dose EDC ligation results in minor changes to the electron density for CC1 with 3′ phosphates, (**F**) high dose EDC ligation results in electron density consistent with ligated DNA. Neighboring protein is hidden for clarity. All meshes are omit maps (mF_o-DF_c) contoured at 3.0 rmsd. All four bases flanking the junction (orange sticks) were omitted. To faithfully represent COOT contours in PyMOL, we turned off automatic map normalization and instead set the contour level to 3.0 rmsd. Table S4 has the corresponding e/Å^3 values.

Consistent with the lower distance between C5′ and O3′ for chain B (Table 3), the electron density was invariably higher for the right hand nick (chain B:chain B). When contoured at 3.0 rmsd, the omit map electron density was even contiguous for the non-ligated CC1 3′P case (Figure 5C). Notably, the maps for crystals subjected to EDC (Figure 5E,F) are discovery maps in the sense that the models were refined in the absence of terminal 3′ phosphates. After light ligation (Figure 5E), the omit map was not clearly changed. However, after heavy ligation (Figure 5F), there was very strong electron density on the right and solid electron density in the left. Phosphates were added prior to submission to the PDB (entry 7spm) and our final refinement calculation for CC1 3′P High included bond length restraints between the model and its symmetry neighbor to ensure a reasonable phosphate geometry.

It is somewhat remarkable that ligation was visible in the electron density trend (Figure 5C–F), despite the incomplete ligation yield suggested by the electrophoresis data (Table 4). In principle, the clarity of the ligation sites in the electron density maps may vary depending on whether the X-ray beam is diffracting from a highly ligated region of the crystal.

3.4. Co-Crystal Stabilization Effects from Ligation and Crosslinking

To determine if crosslinked co-crystals may be suitable for various applications, including biomedical applications at physiologically relevant conditions, the co-crystals were crosslinked (20 h, 15 mg/mL EDC) and subjected to a panel of harsh conditions: a stomach acid mimic, a lysosomal fluid mimic, blood serum (bovine calf), and deionized water (Figure 6). The conditions chosen, especially the stomach acid mimic and deionized water, were challenging for native crystals (no crosslink) since DNA-containing crystals typically require stabilizing counterions.

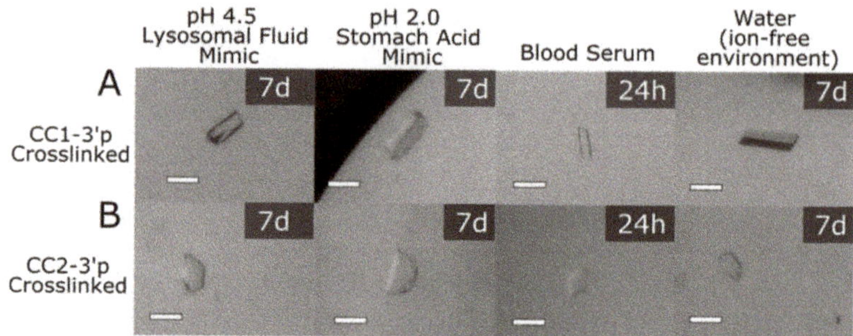

Figure 6. A survey of crosslinked crystals (15 mg/mL EDC 20 h) with terminal 3′ phosphates in four stringent solutions. (**A**) CC1-3′p crosslinked crystals incubated in pH 4.5, pH 2.0 and water for seven days and blood serum for twenty-four hours. (**B**) CC2-3′p crosslinked crystals incubated in pH 4.5, pH 2.0 and water for seven days and blood serum for twenty-four hours.

In the stomach acid mimic (0.01 M hydrochloric acid pH 2), the non-crosslinked co-crystals were observed to convert to an aggregate (Figure S4). Remarkably, in the stomach acid solution, the entire set of crosslinked crystals demonstrated enhanced stability, not dissolving even after 7 days. The 3′ phosphate crosslinked crystals did not change macrostructure for at least 5 days in the harshly acidic environment (Figure S4). Co-crystals without phosphates were also crosslinked and these crystals expanded dramatically in the acidic environment after 24 h (~430 ± 70% volume change), demonstrating the importance of the DNA ligation for crystal stability. Crosslinked co-crystals also maintained integrity in a lysosomal mimic buffer (pH 4.5) and blood serum with no measurable changes to the crystal dimensions after 24 and 72 h, respectively (Figures S5 and S6).

In deionized water, the co-crystal stability resulting from crosslinking was exceptional (Figures 7 and S3). Within one minute of transferring co-crystals to deionized water, non-crosslinked crystals (except for interesting exception CC2-3′P) completely dissolved or were converted to an aggregate. When the co-crystals were crosslinked (20 h, 15 mg/mL EDC), the crystals remained intact and lacked observable changes to their surface quality or dimensions for at least 7 days (Figures 7 and S3). Interestingly, crosslinked co-crystals without terminal phosphates remained unperturbed, just like the 3′ and 5′ phosphorylated crystals. These results indicate that the protein–protein crosslinks created within the co-crystals were sufficient to maintain macroscopic crystal integrity in water. The distinct stability of crosslinked crystals in water confirmed our hypothesis that crystals can be stabilized with new covalent crosslinks. Specifically, the non-covalent interactions that make up crystals can be stabilized with chemical crosslinking and prevent crystals from degrading rapidly in an ion-environment (deionized water).

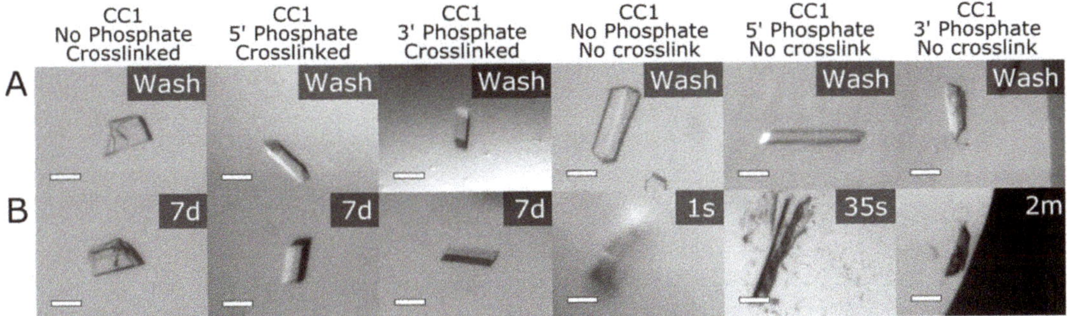

Figure 7. The crystals were crosslinked with 15 mg/mL EDC for 20 h and quenched with Tris base pH 8.2 for 30 min prior to transfer to the wash solution. All scale bars are 100 μm. (**A**) CC1 crystals in wash solution containing 50 mM NaCl, 14% PEG 400, and 200 mM MES buffer pH 6.0. The concentrations of the wash solution matched the initial crystal growth solutions, but we replaced $MgCl_2$ with NaCl and Tris HCl pH 8.0 with MES buffer pH 6.0. (**B**) CC1 crystals after transitioning to an ion-free environment (deionized water). The crosslinked crystals (left three panels) remained intact for 7 days. Non-crosslinked control crystals (right three columns) dissolved or converted to an aggregate at various immediate time points.

4. Discussion

Our strategy in this work was to identify an EDC ligation protocol (EDC concentration, incubation time, and repeated dosage regimen) that optimized reaction yield, without chasing diminishing returns. Accordingly, our final protocol uses 30 mg/mL EDC, an incubation time of 12 h, and two repeated doses within reaction sizes of approximately 200 microliters to ligate the DNA present within approximately 500 ng of co-crystals. Under these conditions, stacked DNA within co-crystals was reliably ligated to a significant extent. We used gel densitometry and detailed Gaussian peak fitting to estimate the fraction of the population for each ligated species (Figure S2).

Global analysis of the ligation product distribution suggested that the most thoroughly crosslinked CC1 crystals feature ligation of nearly half of all possible ligation sites, covalently linking about 50% of the DNA–DNA junctions. Ligation was corroborated by single-crystal XRD where we could directly observe ligation in electron density omit maps (Figure 5).

Apart from small systematic deviations, the random ligation model (RLM, Protocol S2) was able to fit the ligation product distribution for CC1 (Figure S12). In contrast, the CC2 ligation results could not be fit to the RLM as accurately (Figure S12). In particular, the CC2 crystals appeared to have a 1-mer mole fraction that was significantly larger than the total PDSB, which is inconsistent with the RLM. This could be explained by invoking transport limitations. Specifically, one way to boost the 1-mer mole fraction is if the exterior of the crystal has a higher ligation probability than the interior (Protocol S3).

Previously mentioned in the introduction, EDC ligation of DNA has been reported in the literature in the context of DNA hairpins in solution, liquid DNA crystals, and DNA origami. Notably, there has not been a consensus for whether 5′ or 3′ phosphate placement results in a superior yield. Fraccia et al. used 3′ phosphates for the EDC ligation of liquid DNA crystals (15), whereas Kramer and Richert used 5′ phosphates for the EDC ligation of a DNA origami structure (14). Giving a comparison of 5′ versus 3′ phosphates, Obianyor et al. showed EDC ligation of a hairpin DNA structure and reported 95% ligation yield for DNA with 3′ phosphates whereas the 5′ phosphates yielded 40% ligation [17]. They hypothesized the 3′ phosphate ligation reaction could benefit from a primary alcohol nucleophile (Figure 1B) and the geometry difference of the two phosphate positions could contribute to reaction yields. Our data suggest that 3′ phosphates may be superior in the context of a crystal, though comparison between CC1 and CC2 suggests that the results may be system dependent. Our XRD data (Figure 5) furthermore suggest that the results may vary for different nick sites within the same crystal.

It is not clear why 3′ phosphates were more readily ligated than 5′ phosphates in CC1. Conceivably, the rate limiting step for the ligation reaction may be the attack of the hydroxyl on the activated EDC intermediate. Perhaps the short-arm 3′-EDC intermediate is more accessible to the long-arm 5′ hydroxyl than a long-arm 5′-EDC intermediate is to a short-arm 3′ hydroxyl. Notably, one of the 3′ phosphates in the CC1 lattice (chain B) is close (5.65 Å) to a symmetry copy of itself (Figure S13), whereas the 5′ phosphate is farther (9.29 Å). Therefore, 3′ phosphate ligation might be favored due to the greater reduction in electrostatic repulsion upon ligation. However, the CC2 ligation results were more balanced (albeit still favoring 3′ phosphates), suggesting that the relative efficacy of 3′ or 5′ phosphates will be system dependent.

Analyses via gel electrophoresis showed that the ligation yield increased concomitant with the EDC incubation time, but also that the reaction yield appeared to plateau short of full ligation. The cause is unclear. Transport considerations and EDC conjugation to protein sites complicate reaction modeling. One consideration is that the predicted active half-life for EDC in water at 298 K is sixteen hours [17]. However, the ligation yields also appeared to plateau for repeated EDC dosing. Perhaps incomplete ligation is due to a small fraction of DNA strands lacking the necessary terminal phosphate. Alternately, perhaps some EDC-activated phosphates have been ligated to third-party molecules. Perhaps a small DNA strand population is missing a base. Further investigation may be worthwhile prior to future work that depends on near 100% ligation.

In addition to optimizing conditions for our two co-crystals, we have established a set of generalizable guidelines for DNA ligation within co-crystals regarding optimal reaction conditions, phosphate composition, and concentration of EDC. First, it is imperative to optimize the wash solution for each respective system, eliminating components that could interfere with crosslinking. Reactive amines and carboxylic acids are obvious components to eliminate, to avoid forming off-target species. Additionally, we empirically found that it was important to minimize the concentration of the standard divalent cation Mg(II). Re-introducing 90–110 mM Mg(II) into our optimized protocol, we observed a dramatic reduction in the yield (Figure S11). Second, since we found that the best phosphate for ligation may depend on subtle geometry differences, we recommend testing both 5′ and 3′ phosphates for new co-crystal systems. Finally, the EDC concentration used for ligation of a new co-crystal may need to be optimized. Our co-crystals did not dissolve when introduced to crosslinking agents, with the highest concentration at 80 mg/mL. However, in past experiments, we found that the concentration of EDC in the crosslinking reaction drop can affect the integrity of co-crystals. Biomolecular crystals are typically fragile, and a drastic change in solution conditions can cause crystals to fall apart. Therefore, when working with a new system, we recommend testing a range of EDC concentrations. Dosing experiments may be necessary for systems that need a "gentle", multistep transition to harsher conditions. These guidelines may apply to crystals composed of only DNA, as well.

With data for two example co-crystals, generalization is difficult. CC1 and CC2 differ in numerous ways (e.g., DNA length of 21 bp vs. 15 bp, crystal space group, different base pairs spanning the DNA–DNA junction, different DNA sequences in general including flanking base pairs) which makes it difficult to determine which variables may be predictive of ligation yield. Given our observation that ligation may be very sensitive to the nick geometry (Table 3 and Figure 5), we hypothesize that several factors will be particularly important due to their influence on the nick geometry. The DNA sequence at the junction, and to a lesser extent the flanking bases, will affect the base pair stacking energy, which would be expected to change the nick geometry probability distribution. Other nick-site ligation yield differences may be driven by the crystallographic symmetry, particularly the presence or absence of neighboring groups in addition to intrinsic geometry differences between the nick site (e.g., a slightly higher nick distance for chain A nick sites in CC1 crystals).

The crystal stability produced after the chemical ligation of stacked DNA within crystals opens the door for downstream applications, especially for DNA nanotechnology efforts. As shown here, even incomplete ligation can result in dramatic stabilization effects with tangible benefits to suitable application targets. No obvious EDC-induced crosslinks were visible at the two distinct protein–protein interfaces in the CC1 system. Further experiments will be needed to specifically seek and identify any EDC-induced protein–protein or DNA–protein conjugation.

It is possible that DNA ligation provided strong stabilizing effects because both CC1 and CC2 are held together by DNA–DNA junctions in two dimensions (Figures 6 and 7). Essentially, by ligating the stacked DNA in these cases we are forming longer "threads" that are woven together. Stabilization of devices or materials is intriguing if this stabilization allows them to provide or preserve functionality in various biomedical contexts (e.g., in the digestive system, the blood stream, or within lysosomes). It may also be useful if crosslinking allows crystals to remain stable and diffract to high resolution under buffer conditions that mimic physiological conditions (e.g., inside the nucleus), thereby allowing XRD structure determination under conditions besides the idiosyncratic conditions that allow for co-crystal growth.

Along the same lines, one traditional concern crystallographers have regarding crosslinking chemistry is that subjecting a crystal to handling, buffer changes, and reactive chemicals, can degrade the diffraction resolution. For example, subjecting crystals to the common crosslinking agent, glutaraldehyde, can rapidly degrade diffraction resolution. However, supplying aldehydes via gentle vapor diffusion [9] can improve outcomes. We have observed that using glyoxal and EDC can likewise result in negligible diffraction loss, particularly if the reactive chemistry is quenched [11,12]. In the case of CC1, we have once again found that carefully optimized crosslinking protocols can maintain diffraction. Another notable benefit of the EDC crosslinking method is that crystals were not "damaged" during the reaction chemistry. For comparison, when crosslinking HEWL crystals with glutaraldehyde, careful optimization was required to avoid forming cracks in the crystals [31].

Future work may determine if the ligation yield differs for sticky overhang junctions compared to the blunt end junctions used in this work. Similarly, yield may also depend on the DNA bases that span the junction. That said, the current work suggests that the method may be sequence independent because the CC1 junction has a GC/CG and the CC2 junction has an AT/TA. In summary, the reported protocol is a reliable crosslinking strategy using the zero-length crosslinking agent EDC to affect DNA ligation at blunt-end DNA–DNA junctions held together by the co-crystal lattice. Post-ligation stability paves the way for biomedical applications.

Supplementary Materials: The following supporting information can be downloaded at: https://www.mdpi.com/article/10.3390/cryst12010049/s1, Figure S1. TapeStation analysis and matching gel electrophoresis, Figure S2. Densitometry results and annotation (corresponds to main text Figure 4), Figure S3. Co-crystal stability test—water, Figure S4. Co-crystal stability test—very low pH 2.0 to mimic stomach acid, Figure S5. Co-crystal stability test—moderately low pH 4.5 to mimic lysosomal fluid, Figure S6. Co-crystal stability test—blood serum, Figure S7. Gel electrophoresis of varied EDC crosslink time, Figure S8. Gel electrophoresis of varied EDC crosslink concentration, Figure S9. Schematic and gel electrophoresis of varied EDC crosslink dose, Figure S10. Schematic and gel electrophoresis of the controls—crystals with no terminal phosphates and duplexes with terminal phosphates in-solution, Figure S11. Magnesium chloride's effect on the EDC crosslinking of CC1 crystals, Figure S12. Best fits of random ligation model (RLM) to product distribution data, and Figure S13. Terminal phosphates position due to crystallographic symmetry; Table S1. DNA oligonucleotide sequences used in this study, Table S2. Ligation percentages from gel densitometry (unweighted), Table S3. Full version of densitometry output Table 2, and Table S4: Absolute electron density values for the Figure 5 electron density maps; Protocol S1. Protein sequences for cloning and overexpression in *E. coli.*, Protocol S2. Random ligation model: simulation and calculations, Protocol S3. Spatial biased random ligation model, and Protocol S4. Crystal measurements

Author Contributions: Conceptualization, A.R.W., S.D., A.V. and C.D.S.; data curation, A.R.W., S.D., A.V. and C.D.S.; formal analysis, A.R.W., S.D., A.V. and C.D.S.; funding acquisition, A.R.W. and C.D.S.; investigation, A.R.W., S.D., A.V. and C.D.S.; methodology, A.R.W., S.D., A.V. and C.D.S.; project administration, A.R.W. and C.D.S.; resources, C.D.S.; software, A.R.W. and C.D.S.; supervision, A.R.W. and C.D.S.; validation, A.R.W., S.D., A.V. and C.D.S.; visualization, A.R.W. and C.D.S.; writing—original draft, A.R.W. and C.D.S.; writing—review and editing, A.R.W., S.D., A.V. and C.D.S. All authors have read and agreed to the published version of the manuscript.

Funding: This material is based upon work supported by the National Science Foundation under Grant No. NSF DMR 2003748 and NSF DMR 1506219. The team also gratefully acknowledges support for undergraduate researchers from the Nelson Family Faculty Excellence Award.

Institutional Review Board Statement: Not applicable.

Informed Consent Statement: Not applicable.

Data Availability Statement: The data presented in this study are openly available in Zenodo at doi:10.5281/zenodo.5748969.

Acknowledgments: Hataichanok (Mam) Scherman, Histone Source at Colorado State University for the expression and purification of the RepE54 transcription factor and the purification of E2F8 transcription factor. Mark Stenglein and Mikaela Samsel at the Next Generation Sequencing Facility at Colorado State University for TapeStation analysis. Jay Nix at the ALS Beamline 4.2.2 for extensive support of the XRD data collection. The Taipale Lab for their CC2 protein plasmid donation. Thaddaus Huber for cloning expertise and PSB3 plasmid.

Conflicts of Interest: The authors declare no conflict of interest.

References

1. Paukstelis, P.J.; Seeman, N.C. 3D DNA Crystals and Nanotechnology. *Crystals* **2016**, *6*, 97. [CrossRef]
2. Hartje, L.F.; Snow, C.D. Protein Crystal Based Materials for Nanoscale Applications in Medicine and Biotechnology. *Wiley Interdiscip. Rev. Nanomed. Nanobiotechnol.* **2018**, *11*, e1547. [CrossRef]
3. Ward, A.R.; Snow, C.D. Porous Crystals as Scaffolds for Structural Biology. *Curr. Opin. Struct. Biol.* **2020**, *60*, 85–92. [CrossRef] [PubMed]
4. Seeman, N.C. Nucleic Acid Junctions and Lattices. *J. Theor. Biol.* **1982**, *99*, 237–247. [CrossRef]
5. Maita, N. Crystal Structure Determination of Ubiquitin by Fusion to a Protein That Forms a Highly Porous Crystal Lattice. *J. Am. Chem. Soc.* **2018**, *140*, 13546–13549. [CrossRef] [PubMed]
6. Krauss, I.R.; Merlino, A.; Vergara, A.; Sica, F. An Overview of Biological Macromolecule Crystallization. *Int. J. Mol. Sci.* **2013**, *14*, 11643–11691. [CrossRef] [PubMed]
7. Hollis, T. Crystallization of Protein-DNA Complexes. In *Macromolecular Crystallography Protocols*; Humana Press: Totowa, NJ, USA, 2007; Volume 363, pp. 225–237. [CrossRef]
8. Hermanson, G.T. *Bioconjugate Techniques*; Academic Press: Cambridge, MA, USA, 2013; ISBN 978-0-12-382240-6.
9. Lusty, C.J. A Gentle Vapor-Diffusion Technique for Cross-Linking of Protein Crystals for Cryocrystallography. *J. Appl. Crystallogr.* **1999**, *32*, 106–112. [CrossRef]
10. Yan, E.-K.; Cao, H.-L.; Zhang, C.-Y.; Lu, Q.-Q.; Ye, Y.-J.; He, J.; Huang, L.-J.; Yin, D.-C. Cross-Linked Protein Crystals by Glutaraldehyde and Their Applications. *RSC Adv.* **2015**, *5*, 26163–26174. [CrossRef]
11. Hartje, L.F.; Bui, H.T.; Andales, D.A.; James, S.P.; Huber, T.R.; Snow, C.D. Characterizing the Cytocompatibility of Various Cross-Linking Chemistries for the Production of Biostable Large-Pore Protein Crystal Materials. *ACS Biomater. Sci. Eng.* **2018**, *4*, 826–831. [CrossRef]
12. Huber, T.R.; Hartje, L.F.; McPherson, E.C.; Kowalski, A.E.; Snow, C.D. Programmed Assembly of Host–Guest Protein Crystals. *Small* **2017**, *13*, 1602703. [CrossRef]
13. Zhang, D.; Paukstelis, P.J. Enhancing DNA Crystal Durability through Chemical Crosslinking. *ChemBioChem* **2016**, *17*, 1163–1170. [CrossRef] [PubMed]
14. Abdallah, H.O.; Ohayon, Y.P.; Chandrasekaran, A.R.; Sha, R.; Fox, K.R.; Brown, T.; Rusling, D.A.; Mao, C.; Seeman, N.C. Stabilisation of Self-Assembled DNA Crystals by Triplex-Directed Photo-Cross-Linking. *Chem. Commun.* **2016**, *52*, 8014–8017. [CrossRef]
15. Gerling, T.; Kube, M.; Kick, B.; Dietz, H. Sequence-Programmable Covalent Bonding of Designed DNA Assemblies. *Sci. Adv.* **2018**, *4*, eaau1157. [CrossRef]
16. Li, Z.; Liu, L.; Zheng, M.; Zhao, J.; Seeman, N.C.; Mao, C. Making Engineered 3D DNA Crystals Robust. *J. Am. Chem. Soc.* **2019**, *141*, 15850–15855. [CrossRef]

17. Obianyor, C.; Newnam, G.; Clifton, B.; Grover, M.A.; Hud, N.V. Impact of Substrate-Template Stability, Temperature, Phosphate Location, and Nick-Site Base Pairs on Non-Enzymatic DNA Ligation: Defining Parameters for Optimization of Ligation Rates and Yields with Carbodiimide Activation. *bioRxiv* **2019**, 821017. [CrossRef]
18. Fraccia, T.P.; Smith, G.P.; Zanchetta, G.; Paraboschi, E.; Yi, Y.; Walba, D.M.; Dieci, G.; Clark, N.A.; Bellini, T. Abiotic Ligation of DNA Oligomers Templated by Their Liquid Crystal Ordering. *Nat. Commun.* **2015**, *6*, 6424. [CrossRef]
19. Kramer, M.; Richert, C. Enzyme-Free Ligation of 5′-Phosphorylated Oligodeoxynucleotides in a DNA Nanostructure. *Chem. Biodivers.* **2017**, *14*, 1700315. [CrossRef]
20. Komori, H.; Matsunaga, F.; Higuchi, Y.; Ishiai, M.; Wada, C.; Miki, K. Crystal Structure of a Prokaryotic Replication Initiator Protein Bound to DNA at 2.6 A Resolution. *EMBO J.* **1999**, *18*, 4597–4607. [CrossRef] [PubMed]
21. Bi, S.; Pollard, A.M.; Yang, Y.; Jin, F.; Sourjik, V. Engineering Hybrid Chemotaxis Receptors in Bacteria. *ACS Synth. Biol.* **2016**, *5*, 989–1001. [CrossRef] [PubMed]
22. Morgunova, E.; Yin, Y.; Jolma, A.; Dave, K.; Schmierer, B.; Popov, A.; Eremina, N.; Nilsson, L.; Taipale, J. Structural Insights into the DNA-Binding Specificity of E2F Family Transcription Factors. *Nat. Commun.* **2015**, *6*, 10050. [CrossRef] [PubMed]
23. Ward, A.R.; Snow, C.D. *Scripts for Modeling Chemical Ligation of DNA Junctions within Biomolecular Crystals*; Zenodo: Geneva, Switzerland, 2021.
24. Newville, M.; Stensitzki, T.; Allen, D.B.; Ingargiola, A. *LMFIT: Non-Linear Least-Square Minimization and Curve-Fitting for Python*; Zenodo: Geneva, Switzerland, 2014.
25. Kabsch, W. XDS. *Acta Crystallogr. D Biol. Crystallogr.* **2010**, *66*, 125–132. [CrossRef] [PubMed]
26. Afonine, P.V.; Grosse-Kunstleve, R.W.; Echols, N.; Headd, J.J.; Moriarty, N.W.; Mustyakimov, M.; Terwilliger, T.C.; Urzhumtsev, A.; Zwart, P.H.; Adams, P.D. Towards Automated Crystallographic Structure Refinement with Phenix. *Refine. Acta Crystallogr. D Biol. Crystallogr.* **2012**, *68*, 352–367. [CrossRef] [PubMed]
27. Emsley, P.; Lohkamp, B.; Scott, W.G.; Cowtan, K. Features and Development of Coot. *Acta Crystallogr. D Biol. Crystallogr.* **2010**, *66*, 486–501. [CrossRef] [PubMed]
28. Gelbin, A.; Schneider, B.; Clowney, L.; Hsieh, S.-H.; Olson, W.K.; Berman, H.M. Geometric Parameters in Nucleic Acids: Sugar and Phosphate Constituents. *J. Am. Chem. Soc.* **1996**, *118*, 519–529. [CrossRef]
29. Lu, X.-J.; Olson, W.K. 3DNA: A Versatile, Integrated Software System for the Analysis, Rebuilding and Visualization of Three-Dimensional Nucleic-Acid Structures. *Nat. Protoc.* **2008**, *3*, 1213–1227. [CrossRef]
30. Olson, W.K.; Gorin, A.A.; Lu, X.-J.; Hock, L.M.; Zhurkin, V.B. DNA Sequence-Dependent Deformability Deduced from Protein–DNA Crystal Complexes. *Proc. Natl. Acad. Sci. USA* **1998**, *95*, 11163–11168. [CrossRef]
31. Yan, E.-K.; Lu, Q.-Q.; Zhang, C.-Y.; Liu, Y.-L.; He, J.; Chen, D.; Wang, B.; Zhou, R.-B.; Wu, P.; Yin, D.-C. Preparation of Cross-Linked Hen-Egg White Lysozyme Crystals Free of Cracks. *Sci. Rep.* **2016**, *6*, 34770. [CrossRef]

Article

A New L-Proline Amide Hydrolase with Potential Application within the Amidase Process

Sergio Martinez-Rodríguez [1,2,*], Rafael Contreras-Montoya [3], Jesús M. Torres [1], Luis Álvarez de Cienfuegos [3] and Jose Antonio Gavira [2,*]

1 Department of Biochemistry and Molecular Biology III and Immunology, University of Granada, 18071 Granada, Granada, Spain; torrespi@ugr.es

2 Laboratory of Crystallographic Studies, Andalusian Institute of Earth Sciences, C.S.I.C. University of Granada, Avenida de las Palmeras No. 4, 18100 Armilla, Granada, Spain

3 Department of Organic Chemistry, University of Granada, 18071 Granada, Granada, Spain; rcm@ugr.es (R.C.-M.); lac@ugr.es (L.Á.d.C.)

* Correspondence: sergio@ugr.es (S.M.-R.); jgavira@iact.ugr-csic.es (J.A.G.)

Abstract: L-proline amide hydrolase (PAH, EC 3.5.1.101) is a barely described enzyme belonging to the peptidase S33 family, and is highly similar to prolyl aminopeptidases (PAP, EC. 3.4.11.5). Besides being an *S*-stereoselective character towards piperidine-based carboxamides, this enzyme also hydrolyses different L-amino acid amides, turning it into a potential biocatalyst within the Amidase Process. In this work, we report the characterization of L-proline amide hydrolase from *Pseudomonas syringae* (PsyPAH) together with the first X-ray structure for this class of L-amino acid amidases. Recombinant PsyPAH showed optimal conditions at pH 7.0 and 35 °C, with an apparent thermal melting temperature of 46 °C. The enzyme behaved as a monomer at the optimal pH. The L-enantioselective hydrolytic activity towards different canonical and non-canonical amino-acid amides was confirmed. Structural analysis suggests key residues in the enzymatic activity.

Keywords: amidase; amino acid; amidase process; proline; aminopeptidase; S33 family

Citation: Martinez-Rodríguez, S.; Contreras-Montoya, R.; Torres, J.M.; de Cienfuegos, L.Á.; Gavira, J.A. A New L-Proline Amide Hydrolase with Potential Application within the Amidase Process. *Crystals* **2022**, *12*, 18. https://doi.org/10.3390/cryst12010018

Academic Editors: Kyeong Kyu Kim and Dinadayalane Tandabany

Received: 25 November 2021
Accepted: 21 December 2021
Published: 23 December 2021

1. Introduction

L-proline amide hydrolase (PAH, EC 3.5.1.101) is a barely described enzyme, which up to now, has only been characterized with some detail in *Pseudomonas azotoformans* IAM 1603 (LaaA$_{Pa}$) [1,2]. PAH belongs to the serine peptidase S33 family, together with prolyl aminopeptidases (PAP, EC. 3.4.11.5) or prolinases (Pro-Xaa dipeptidase, 3.4.13.18). PAH was suggested as a different member of this family since LaaA$_{Pa}$ proved a different substrate scope than PAPs [2]. On the other hand, the enzyme proved enantioselective towards different piperidine-based carboxamides, L-prolinamide, and other different amino acid amides (Figure 1A). Since LaaA$_{Pa}$ was applied in the context of the so-called "Amidase Process" for the industrial production of optically pure amino acids, its different substrate scope prompted its nomenclature also as L-amino acid amidase [2–4]. This biotechnological process consists of the dynamic kinetic resolution of amino acid amides mixtures using an α-amino-ε-caprolactam racemase together with a stereoselective "D- or L-amidase" (Figure 1B, [3]).

As for other enzymes with biotechnological interest, the general "amidase" nomenclature might confuse neophyte and experienced researchers, since it includes different unrelated enzymes. The enzymatic resolution of the two isomers of proline amide (D and L) was already achieved using an "amidase" from hog kidney more than half a century ago [5]; this enzyme also proved useful for the resolution of diverse amino acid amides ([6] and references therein).

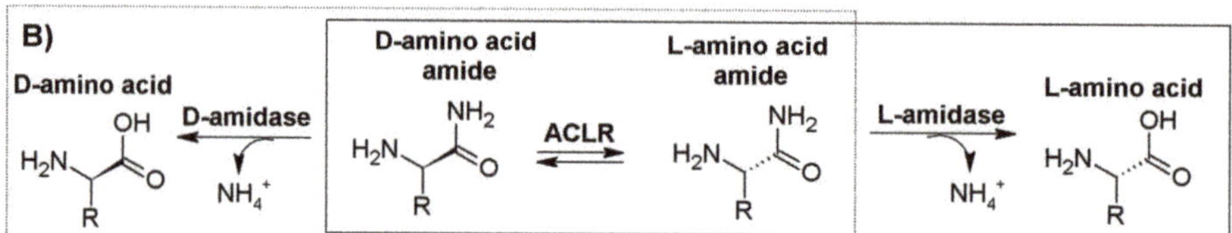

Figure 1. (**A**) Substrates recognized by L-proline amide hydrolase. (**B**) General scheme of the "Amidase Process". The full line represents the "L-system", whereas the dashed line represents the "D-system". ACLR: α-amino-ε-caprolactam racemase.

A reduced number of "L-amidases" have been studied to some detail, such as those from *Pseudomonas azotoformans* [1], *Ochrobactrum anthropi* [7,8], and *Brevundimonas diminuta* [2]. Enzymes from *Pseudomonas putida* [9] or *Mycobacterium neoaurum* ATCC 25795 [10], and different aminopeptidases (EC. 3.4.11.X) and amidases (E.C. 3.5.1.4) have also shown to able to hydrolyze amino acid amides with good enantioselectivity [11,12]. Some of the latter enzymes have been applied at the industrial level [9,11,12]. A discrete example of the hydrolysis of amino acid esters and amides by acylase I has been reported, despite this enzyme being mainly used for the hydrolysis of N-acetyl-amino acids [13]. Peptide amidase from *Citrus sinensis* and *Stenotrophomonas maltophilia* also allowed enzymatic resolution of racemic N-acetyl amino acid amides, yielding N-acetyl-L-amino acids with optical purity $\geq$ 99% [14]. As for the "amidase" nomenclature, PAPs present a similar scenario, whereas many of the reported PAPs show a clear preference for proline residues ([15] and references therein), not all cases show that they are obligate "proline aminopeptidases". Some members of this family have shown cleaving activity with different amino acid derivatives at different extents [15–17].

In order to gain understanding into enzymes with L-amidase activity and with potential industrial interest, we have embarked on the characterization of a putative PAP from *Pseudomonas syringae* (PsyPAH). This enzyme is highly similar to $LaaA_{Pa}$, which is the only PAH characterized showing L-amino acid amidase activity [1]. On the other hand, the closest structural homolog of PsyPAH to date is the amidohydrolase VinJ from *Streptomyces halstedii* (PDB 3WMR, 55% seq id.), with a highly different substrate scope [18]. In this work, we provide biochemical and biophysical characterization, together with the first X-ray structure of a PAH enzyme (PAP-like) with experimentally proven "L-amidase" activity. We have gone a step forward and based on sequence and structural information, we have categorized the different L-amidase enzymes in the literature in an attempt to facilitate comprehension on their potential biotechnological application.

2. Materials and Methods

The different amino acid amides and *p*-nitroanilide derivatives used for activity measurement of PsyPAH were purchased from VWR (VWR International Eurolab S.L, Barcelona, Spain), TCI chemicals, Alfa Aesar, or Acros (Cymit Quimica, Barcelona, Spain). Other amino acid amides were synthesized as previously described [19] (see supporting information). Other chemicals were from Sigma Aldrich (Sigma-Aldrich, St. Louis, MO, USA).

2.1. Cloning, Overexpression, and Purification of PsyPAH

A DNA sequence corresponding to the putative L-amidase from *Pseudomonas syringae pv. tomato* (Uniprot A0A0Q0CYJ4) was synthesized and cloned into pET-22b (NZYtech, Lisboa, Portugal) for over-expression in *Escherichia coli*. The resulting construct allows the overproduction of PsyPAH fused to a C-terminal His_6-tag. *E. coli* BL21 (DE3) (Agilent, Madrid, Spain) was transformed with this plasmid and grown in solid LB medium supplemented with 100 μg·mL^{-1} of ampicillin. A single colony was transferred into 10 mL of LB medium with ampicillin at the concentration above mentioned and incubated overnight at 37 °C. Then, 500 mL of LB supplemented with ampicillin was inoculated with 5 mL of the overnight culture. After 3–4 h of incubation at 37 °C with vigorous shaking, the OD^{600} of the resulting culture was 0.6–0.8. To induce the over-expression of PsyPAH, isopropyl-β-thio-D-galactopyranoside (IPTG) was added to a final concentration of 0.2 mM and the culture was kept at 16 °C overnight. Cells were collected by centrifugation (4000 rpm, 4 °C, 20 min) and subsequently frozen at −80 °C till use.

The pellet corresponding to 1 L was resuspended in 10 mL of 20 mM sodium phosphate, 20 mM of imidazole, and 300 mM of NaCl pH 8.0 (washing buffer, WB). Cells were lysed on ice via sonication with a Branson sonicator (6 periods of 60 s (1 s on, 1 s off), amplitude 25%) and then centrifuged (13,000 rpm, 10 min, RT). The resulting supernatant was applied to a HisPur Ni-NTA column (1 mL, Thermo Fisher, Waltham, MA, USA) previously equilibrated with 10 mL of WB. The column was then washed with 12 mL of WB and protein was eluted with 3 mL of 20 mM of sodium phosphate, 300 mM of imidazole, and 300 mM of NaCl pH 8.0. Subsequently, protein samples were loaded onto a Superdex 200 16/60 XK gel-filtration column (GE Healthcare, Boston, MA, USA) in an AKTA-prime FPLC system (GE Healthcare) using 20 mM of Hepes pH 7.0 as a running buffer. The peak corresponding to PsyPAH was concentrated up to 20 mg·mL^{-1} using 30 kDa concentrators (Amicon Ultra-Millipore) and dialyzed in 20 mM of Hepes pH 7.0 (4 °C). Protein was frozen at −80 °C till use. Protein purity was verified by SDS-PAGE. Protein concentrations were determined from the absorbance at 280 nm (ε = 49,390 M^{-1}·cm^{-1}).

2.2. Activity Measurement

Different amino acid amides (10 mM) were used as possible substrates for PsyPAH: (amide derivatives of Gly, L-Pro and D-Pro, L-Trp, L-Phe, L-*tert*-Leu, L-Ala, L-norVal, L-Met, L-homoPhe, L-Ser, L-norLeu, L-Leu, L-2-ABA, and L-Val). The phenate method was used to measure ammonia formation [20], with slight modifications. Reaction volumes of 200 μL and a final enzyme concentration of 0.1–0.2 mg·mL^{-1} (pH 7.0, 35 °C) were used. After 5–15 min, the reaction was stopped by mixing with 540 μL of freshly prepared phenate solution. A total of 280 μL of 2.5% sodium hypochlorite and 140 μL of 25 μM $MnCl_2$ were then added, followed by incubation at 70 °C for 40 min. Absorbance was measured at 625 nm. $(NH_4)_2SO_4$ standards were used for all the assays. Three replicates were conducted for each experiment.

Kinetic parameters for L-prolinamide and L-leucinamide were calculated with substrate concentrations ranging 0.1 to 15 mM, using 100 mM of stock solutions (in 100 mM of phosphate buffer pH 7.0). Reactions were carried out at 35 °C and pH 7.0 (using 20–400 μg·mL^{-1} PsyPAH concentrations depending on the substrate). After 5–15 min, (pre-experiments suggested this reaction time as appropriate for Vo calculation), ammonium formation was measured with the phenate method (see above). The activity with *p*-nitroanilide derivatives was measured following sample absorption at 405 nm. K_m and k_{cat} were measured using L-Leu and L-Pro *p*-nitroanilide concentrations ranging from 0.1 to 10 mM, using 500 mM of stock solutions (in acetonitrile). Reactions were carried out at 35 °C and pH 7.0 (using 8–80 ng·mL^{-1} PsyPAH concentrations depending on the substrate, with a constant 2% concentration of acetonitrile into the reaction). A calibration was performed and plotted with *p*-nitroaniline in the same buffer used for activity determination (experimental ε = 9265 cm^{-1}·M^{-1}, similar to that reported previously [21]). Three replicates were conducted for each experiment.

2.3. *Size Exclusion Chromatography (SEC-FPLC)*

PsyPAH was loaded onto a Tricorn Superdex 200 gel-filtration column (GE Healthcare) using an AKTA-prime FPLC system (GE Healthcare), with 20 mM of sodium phosphate pH 7.0 as a running buffer. BSA (66 kDa), ovalbumin (43 kDa), carbonic anhydrase (29 kDa), and RNase A (13.7 kDa) were used as standards for molecular mass determination (Cytiva Gel Filtration Calibration Kits).

2.4. *Dynamic Light Scattering*

DLS measurements were performed in a Zetasizer Nano instrument (Malvern Instruments Ltd., Malvern, UK). Experiments were performed with PsyPAH (1.3 mg·mL^{-1}) in 20 mM of sodium phosphate pH 7.0 at 25 °C. Samples were centrifuged for 10 min at 13,000 rpm before measurement. The PsyPAH sample was measured 3 times with 10 runs each (in automatic mode for time selection).

2.5. *Thermal Shift Assays*

Thermal shift assays were carried out using a QuantStudio 3 qPCR (Thermo Fisher). A concentrated PsyPAH sample was 10-fold diluted directly into different 100-mM buffers (sodium acetate, pHs 4.0–5.6; sodium phosphate, pHs 6.0–8.0; tetraborate HCl/NaOH, pHs 8.0–10.0) to a final concentration of 1.4 mg·mL^{-1}, and kept at 4 °C O/N. Aqueous SYPRO (50×) was added to a final 10× concentration. Thermal denaturation measurements were monitored by measuring the changes in the fluorescence as a result of SYPRO binding. Denaturation data were collected from 25 to 99 °C at a scan rate of 3 °C·min^{-1}. Three replicates were conducted in all cases. Despite the irreversibility of the thermal unfolding, apparent Tms were calculated using a Boltzmann fit to the raw data, with Protein Thermal shift software v1.3 (Thermo Fisher).

2.6. *Crystallization*

Freshly purified recombinant His$_6$-tagged PsyPAH (20 mg·mL^{-1}, 20 mM of Hepes pH 7.0) was used to set up initial crystallization screenings with the HRCS I & II (Hampton Research, Palo Alto, CA, USA). The hanging drop configuration of the vapor diffusion method with a 1:1 ratio of the reservoir and protein solution was used. Crystallization experiments were kept at 20 °C in an incubator. Crystals were obtained using 0.2 M of sodium acetate trihydrate, 0.1 M of sodium cacodylate trihydrate pH 6.5, and 30% w/v polyethylene glycol 8000 after 48 h.

2.7. *Data Collection and Refinement*

Target crystals were identified under a microscope using polarized light, separated with a microtool, fished out of the drop with a loop, and transferred to a 1-µL drop of mother solution containing 20% (v/v) glycerol as cryo-protectant. After soaking for less than 60 s, crystals were flash-cooled in liquid nitrogen and stored until data collection.

X-ray diffraction data were collected at ID30B (ESRF, Grenoble, France). Diffraction data were indexed and integrated using XDS [22] and scaled with AIMLESS from the CCP4 suite [23]. The crystal structure of PsyPAH was determined by the molecular replacement method with PHASER [24] using the structure of the amidohydrolase VinJ from *Streptomyces halstedii* (PDB ID: 3WMR) [18] as the search model. Refinement was done with PHENIX [25] and Refmac [26] with cycles of manual rebuilding using COOT [27] and finalized using several cycles of refinement applying TLS parameterization [28]. The final refined model was checked with Molprobity [29]. Data collection and refinement statistics are summarized in Table 1.

Table 1. Data collection and refinement statistics. (Statistics for the highest-resolution shell are shown in parentheses.)

Data Collection	
Source	ESRF ID30B
Space group	P 21 21 21
Cell dimensions	
a, b, c (Å)	49.428, 65.033, 85.016
α, β, γ (°)	90.0, 90.0, 90.0
Resolution (Å)	42.73–1.95 (2.02–1.95)
Unique reflections	20,565 (2016)
$I/\sigma I$	7.2 (1.8)
Completeness (%)	99.70 (99.75)
Redundancy	5.1 (5.1)
R_{merge}	15.7 (91.9)
$CC_{1/2}$	0.991 (0.636)
Refinement	
R_{work}/R_{free}	15.68/22.25
No. atoms	2824
Protein	2571
Ligand/ion	5
Water	248
B-factors	21.97
Protein	20.95
Ligand/ion	30.16
Water	32.35
R.m.s. deviations	
Bond lengths (Å)	0.007
Bond angles (°)	0.85
PDB ID	7A6G

2.8. Sequence and Structure Analysis

PDB-SUM was used for global structure analysis [30]. Clustal omega [31] and SPript [32] were used for multiple sequence alignment and phylogenetic analysis. The i-Tol server was used for tree representation [33]. The Dali server [34] was used to search for other members of the peptidase S33 superfamily with a similar fold to that presented by the PsyPAH structure. Graphical representation of 3D structural models was conducted with Pymol [35].

3. Results and Discussion

3.1. PsyPAH Characterization

Recombinant C-His_6-tagged PsyPAH was purified using nickel affinity chromatography and SEC-FPLC (Size-exclusion chromatography-Fast Protein Liquid Chromatography) (>95% purity, yield of 10 mg per L of culture). SEC-FPLC showed an estimated molecular mass of 33 ± 2 kDa in phosphate buffer pH 7.0 (Figure 2A), slightly lower than the theoretical molecular mass of the monomer (36.7 kDa). An estimated R_h of 2.5 ± 0.40 nm was obtained for PsyPAH by DLS (20 mM phosphate pH 7.0). This value is a bit higher than that shown for carbonic anhydrase (29 kDa, 2.37 nm [36]), and argues with the value obtained by SEC-FPLC. Thermal Shift Assays (TSA) showed single thermal transitions in the pH range from 6.0 to 11.0 as a result of SYPRO binding (Figure 2B, inset).

Apparent thermal midpoints (T_m^{app}) could be calculated from Boltzmann fitting, with values ranging from 35.8 to 46.0 °C in that pH interval (Figure 2B). The maximum T_m^{app} value coincided with the optimum pH activity of the enzyme (pH 7.0; Figure 2C). The optimal reaction temperature was 35 °C (Figure 2D), whereas enzymatic activity was lost at 50 °C.

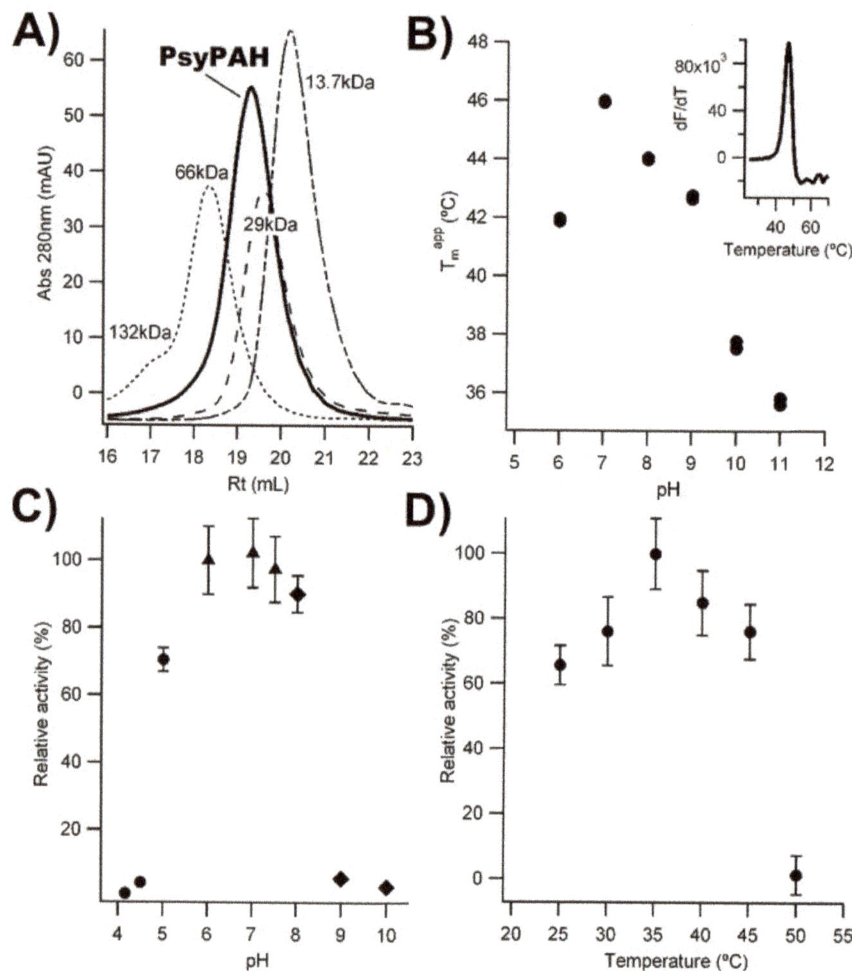

Figure 2. (**A**) SEC-FPLC of PsyPAH (black continuous line) in phosphate buffer 20 mM pH 7.0. Protein standards represented in dashed lines are BSA (132 and 66 kDa), carbonic anhydrase (29 kDa), and RNAse A (13.7 kDa). (**B**) Apparent T_ms calculated for PsyPAH at different pHs. The inset corresponds to the TSA experiment of PsyPAH in phosphate buffer at pH 7.0. Relative activity of PsyPAH as a function of pH (**C**) and Temperature (**D**).

Both the optimal temperature and pH were lower than those reported previously for $LaaA_{Pa}$ [2]. No activity loss was observed after incubation of PsyPAH at 30 °C for 14 h and it also retained over 75% of its activity after incubation at 35 °C for the same period of time. PsyPAH stored at −80 °C maintained full activity for more than two years. Biochemical parameters of PsyPAH were assayed with the amide and *p*-nitroanilide derivatives of L-Pro and L-Leu (Table 2), showing the expected L-amidase activity of the enzyme. Whilst, we could not determine the K_m values for two of the substrates used due to the limit of detection of the method (L-Pro-*p*-nitroanilide) and the solubility of the substrate (L-Leu-amide), visual inspection of the kinetic profiles (Figure S1) supports that the K_m for the amide derivatives of L-Pro and L-Leu is, at least, one order of magnitude higher than for the *p*-nitroanilide derivative (Table 2 and Figure S1). These results suggest that the presence of the aromatic aniline moiety of the substrate improves PsyPAH-binding, which might reflect a better accommodation of these substrates into the active site.

Table 2. Kinetic parameters of PsyPAH with L-Pro and L-Leu amide and *p*-nitroanilide derivatives (pH 7.0, 35 °C). * Could not be determined due to detection limit of the determination method. ** Could not be determined due to the solubility of this substrate. *** Obtained from the linear part of the kinetic plot (see Figure S1).

Substrate	K_m (mM)	k_{cat} (s^{-1})	k_{cat}/K_m ($s^{-1}\cdot mM^{-1}$)
pN-Pro	ND *	445.38 ± 19.17	ND *
pN-Leu	1.14 ± 0.21	71.25 ± 3.34	62.5 ± 3.37
L-Pro amide	9.48 ± 1.6	14.03 ± 1.26	1.48 ± 0.13
L-Leu amide	ND **	ND **	0.03 ± 0.00 ***

We have also qualitatively tested the activity of PsyPAH towards different canonical and non-canonical L-amino acid amides. PsyPAH was able to hydrolyze glycinamide, L-alaninamide, L-phenyalalaninamide, L-methioninamide, L-serinamide, L-valinamide, L-tryptophanamide, L-norvalinamide, L-homophenylalaninamide, L-norleucinamide, and L-2-aminobutyramide (data not shown). No activity was detected towards D-prolinamide or L-tert-leucinamide.

3.2. PsyPAH Sequence Analysis

Since E.C. classification is based solely on the enzymatic reaction, different enzymes catalyzing the same reaction can share the same nomenclature (e.g., L-amidases), even when their sequences are highly different. This is a recurrent issue in the biotechnological field, where it is common to discover novel enzymes after screening methods for a desired specific activity, from which they are named. The general "amidase" nomenclature used in the context of the "Amidase Process" might thus initially confuse neophyte researchers in this field, since many different enzymes classified under E.C. 3.5.1 are named as "amidases" [12,37]. Previous studies on L-amidases of biotechnological interest already highlighted enzymes belonging to different protein families [2,12].

Phylogenetic analysis of the primary sequence of enzymes with L-amidase activity shows four different enzyme groups (Table 3 and Figure S2). The broad-spectrum amidase from *Ochrobactrum anthropi* [7] shapes an alternative "acetamidase/formamidase clan" (Pfam PF03069), together with the enzymes from *Enterobacter cloacae* and *Thermus* sp. (Table 2). The industrially-used L-amidase from *Pseudomonas putida* (a leucine aminopeptidase [9]) and $LaaA_{Bd}$ shape an alternative "aminopeptidase clan", belonging to the peptidase M17 family (Pfam PF00883). On the other hand, the leucyl-aminopeptidase from *Aeromonas proteolytica* [38] confers an isolated clan, which belongs to the peptidase M28 family (Pfam PF04389, Table 3). Finally, $LaaA_{Pa}$ and PsyPAH are grouped into a "peptidase S33 clan".

Thus, from a biotechnological point of view, it is important to bear in mind that different "L-amidases" belonging to different protein families exist when dealing with the so-called "Amidase Process". Besides their application on the production of amino acids, some of these L-amidases have also found other biotechnological applications [37,39,40], further increasing their potential and economic interest.

Table 3. Different enzymes with L-enantioselective amidase activity described in the literature with potential application in the production of amino acids. * It is not clear from the literature whether the hog kidney amidase used in the 50s for the resolution of amino acids [5,6] might correspond to a leucyl aminopeptidase or a PAP, or even if they are the same enzyme [41,42].

Acronym	Protein Family	Source	Sequence	Reference
PsyPAH	Serine peptidase S33	*Pseudomonas psyringae*	A0A0Q0CYJ4	This work
$LaaA_{Pa}$	Serine peptidase S33	*Pseudomonas azotoformans*	BAD15092.1	[1]
$LaaA_{Oa}$	Acetamidase/formamidase	*Ochrobactrum anthropi*	AAV87210	[7]
$LaaA_{Ec}$	Acetamidase/formamidase	*Enterobacter cloacae*	AAR56843	[43]
$LaaA_{Ts}$	Acetamidase/formamidase	*Thermus* sp.	BAL49703.1	[44]
XfAmid	Acetamidase/formamidase	*Xantobacter flavus*	BAE02548.1	[45,46]
$LaaA_{Bd}$	Leucine aminopeptidase (M17)	*Brevundimonas Diminuta*	BAE91931	[2]
ppLAP	Leucine aminopeptidase (M17)	*Pseudomonas putida*	CAA09054.1	[9]
apLAP	Leucyl-aminopeptidase (M28)	*Aeromonas proteolytica*	Q01693	[38]
$LaaA_{Mn}$	-	*Mycobacterium neoaurum*	n.a.	[10]
-	- *	*Hog kidney*	- *	[5]

3.3. *Overall Structure of PsyPAH*

PsyPAH crystallized in the most standard space group P212121 and presents a single polypeptide chain in the asymmetric unit, as observed in the solution. As ascertained from primary sequence analysis, PsyPAH belongs to the hugely diverse α/β hydrolase superfamily and more specifically to the serine peptidase family S33 (clan SC) [47]. The α/β hydrolase fold family of enzymes is one of the largest groups of structurally related enzymes with diverse catalytic functions. It contains several enzymes found to have a second promiscuous function on alternative substrates [48,49]. Like other members of this family, PsyPAH is constituted by two different domains, namely the catalytic domain (residues 1–141 and 231–end) and the cap domain (residues 142–230; Figure 3). The catalytic domain is formed by a $\alpha\beta\alpha$ sandwich containing the conserved catalytic triad motif of the family (Ser113, Asp253, His280), whereas the cap domain is constituted exclusively by α-helices. A DALI search shows more than 140 structures with a Z-score over 20 when using the PDB90 subset database. However, only three structures present a sequence similarity over 25% with PsyPAH (Table S1): The amidohydrolase VinJ from *Streptomyces halstedii* (PDB 3WMR, 55% seq id. [18]), a putative uncharacterized PAP from *Mycobacterium smegmatis* (MysPAP, PDB 3NWO, 50% seq id.), and the Tricorn protease-interacting aminopeptidase F1 from *Thermoplasma acidophilum* (APF1, 34% seq. id., PDB 1MU0, [50], with RMSD of 1.0, 2.2, and 1.7 Å, respectively). Other different peptidase S33 family members appear with sequences below 21%, such as epoxide hydrolases and esterases (Table S1). On the other hand, other characterized PAPs included in the ESTHER database [51] whose structures are known, present a lower structure similarity with PsyPAH. This is the case of *Xanthomonas campestris* PAP (XcPAP, PDB 1AZW, [52]) or *Serratia marcescens* PAP (SmPAP, PDB 1QTR, [53]). Other known PAP family structures are those from the PAP-related protein TTHA1809 from *Thermus thermophilus* (PDB 2YYS, [54]) and putative PAP from yeast *Glaciozyma antarctica* (PDB 5YHP, unpublished results). However, no biochemical data is available for these enzymes.

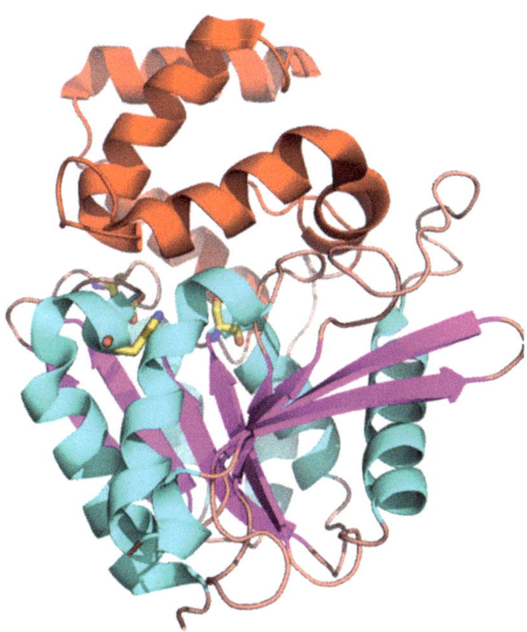

Figure 3. Overall fold of PsyPAH. The catalytic domain is shown in cyan/purple, whereas the cap domain appears in red. The catalytic triad (Ser113, Asp253, His280) is shown in stick mode to account for its position in the catalytic domain.

3.4. Differences on the Substrate Binding Groove (SBG) Seem to Account for the Substrate Scope of PsyPAH

The substrate specificity and function of prolyl peptidases was proposed early on to be determined by the cap domain [53], where the substrate firstly needs to bind before reaching the catalytic center to be hydrolyzed. The specificity of the exopeptidase activity of SmPAP was thus proposed to have originated by steric impediments of this smaller domain, which would block the entrance of extra residues at the N-terminal proline of the substrate [53]. Differences on the substrate binding entrance were already highlighted for APF1, XcPAP, and SmPAP, with the latter showing larger openings to the active site [50]. An overview of the homolog PsyPAH structures reveals that whereas the catalytic domains are spatially conserved, the cap domain presents clear positional differences (Figure 4). Interestingly, the highest differences in the cap domain are observed when comparing PsyPAH with the two characterized PAPs: XcPAP and SmPAP (Figure 4A), while better fit are observed with VinJ, APF1, and the uncharacterized MysPAP (Figure 4B). The substrate binding groove of APF1 (SBG, also known as E1 site [50,55]) was experimentally deciphered between two helices comprised in the cap domain (e.g., PDB 1XRP, Figure 5A); conservation of the spatial disposition of the SBGs into the cap domains of PsyPAH, VinJ, and MysPAP is observed (Figures 5 and S3), revealing clear differences on the different PAP structures: The SBG on XcPAP (and SmPAP) is in a completely different position to the other enzymes, in-between the cap and the catalytic domains (Figure 4A, [53]). The different position of the SBG makes the catalytic center more accessible to the solvent in XcPAP and SmPAP, supporting the acceptance of long peptides. However, in APF1, the N-terminal peptide needs to enter the catalytic center by a narrow hollow, where it can be processed [50,55]. This "smaller" SBG supports the acceptance of shorter peptides when compared with XcPAP and SmPAP. This should also be the case for PsyPAH as observed by the SBG configuration (Figures 4, 5 and S3).

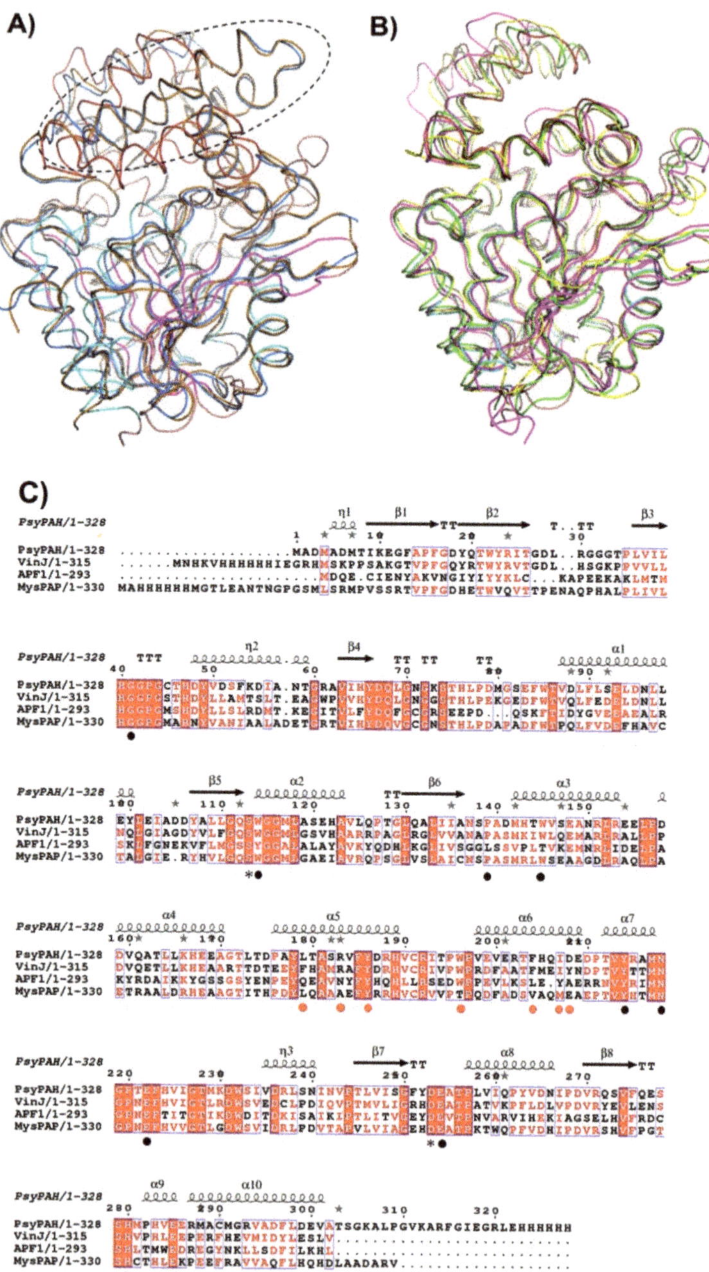

Figure 4. (**A**) Superposition of PsyPAH with "real" PAPs belonging to *Xanthomonas campestris* (XcPAP, PDB 1AZW) and from *Serratia marcescens* (SmPAP, PDB 1QTR). (**B**) Superposition of PsyPAH, amidohydrolase VinJ (PDB 3WMR), APF1 from *Thermoplasma acidophilum* (PDB 1MTZ), and putative uncharacterized PAP from *Mycobacterium smegmatis* (MysPAP, PDB 3NWO). (**C**) Sequence alignment of PsyPAH, VinJ, APF1, and MysPAP shown in panel B. Residues comprising the SBGs (red circles) or catalytic triad (asterisks)/catalytic cleft (black circles) are highlighted.

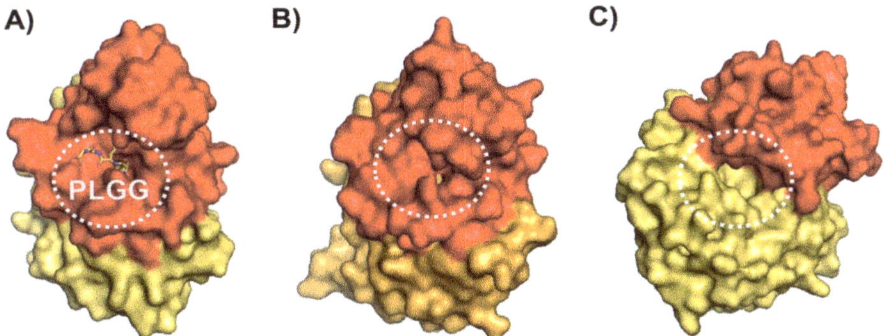

Figure 5. (**A**) Surface representation of APF1 showing the binding site of the tetrapeptide PLGG (E213Q mutant, PDB 1XRP). Red: Cap domain; orange: Catalytic domain. (**B**) Surface representation of PsyPAH showing the putative substrate binding site. Red: Cap domain; orange: Catalytic domain. The orientation is exactly the same as in (**A**). (**C**) Surface representation of XcPAP showing the putative substrate binding site. Red: Cap domain; orange: Catalytic domain. The orientation has been rotated approximately 90° with respect to (**A**).

The closest structural homolog of PsyPAH known to date is VinJ [18], but the most and only exhaustive analysis of the binding mode among homolog structures has been carried out with APF1 [50] (Table S2). Comparison of residues comprising different regions of APF1 (E1, S1, S1′ [50,55]) with those of PsyPAH, VinJ, and MysPAP reveals totally conserved residues, despite a low overall conservation (Table S3). The different substrate scope of VinJ compared to APF1 was explained by the presence of a unique polyketide binding tunnel (which partly correspond to the E1 site, Tables S3 and S4) and a smaller S1 site in VinJ [18]. These unique feature of VinJ is necessary for polyketide moiety fitting on the surface of the enzyme (and other VinJ-proteins used for the synthesis of β-amino acid containing macrolactams [18]). Comparison of this hydrophobic tunnel with PsyPAH, APF1, and MysPAP confirms the unique character of this binding site in VinJ, which shows an overall higher hydrophobic character (Figure S4, Table S4). Specifically, residues $F176^{VinJ}$ and $Y205^{VinJ}$ were hypothesized to provide additional hydrophobic interactions with the polyketide chain of the substrate [18]. However, counterpart residues in PsyPAH, APF1, and MysPAP are overall more polar (Table S4). In this sense, $E200^{APF1}$ (counter part of $Y205^{VinJ}$) has been experimentally proven to be responsible for peptide docking [55], (see below). These structural differences suggest that PsyPAH is not a VinJ-type protein, and also supports a closer binding mode and catalytic mechanism to that reported for APF1 (Table S4).

Different Proline-containing liganded structures of APF1 (Table S2), PDBs 1XQY, 1XRP, and 1XRR, [55] show $Y178^{APF1}$ and $E200^{APF1}$ as responsible for Pro-docking at the E1 site (Figure 6). The counterpart of $Y186^{PsyPAH}$ and $D208^{PsyPAH}$ residues plausibly have a key role in substrate positioning at PsyPAH; in fact, $D208^{PsyPAH}$ shows alternative orientations, suggesting a dynamic character for substrate binding. A lower volume of the PsyPAH SBG is observed when compared to APF1 (Figure 6) arising from (i) displacement of P176-$V190^{PsyPAH}$ helix towards the catalytic domain (originating from the "closure" of the frontside of the E1 site by L179-$D208^{PsyPAH}$) and (ii) the presence of longer or more voluminous side chains at the backside of the E1 site (Figure 6).

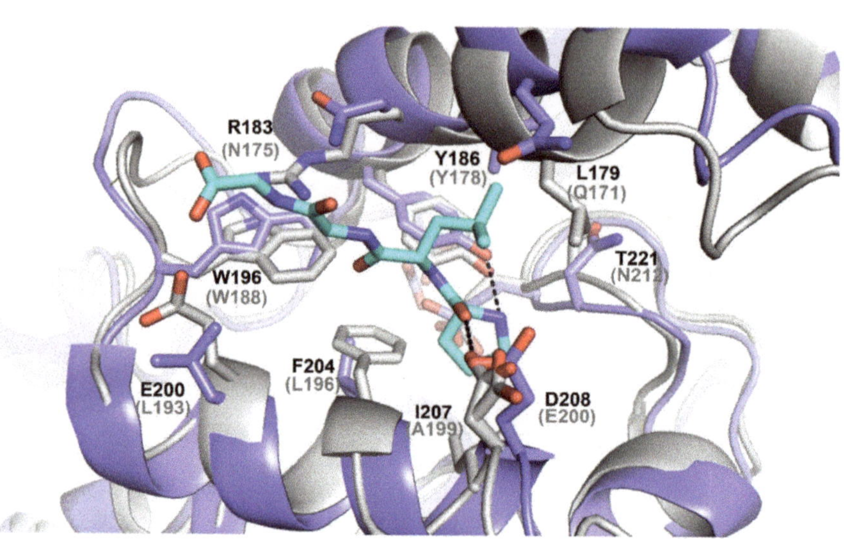

Figure 6. Superposition PsyPAH (white tones) and APF1 bound to PLGG peptide (PDB 1XQY, main chain in blue tones, peptide in sky blue tones). The numbering of the enzymes is that which appears in the corresponding PDBs (PsyPAH, black numbering; APF1, grey numbering).

R183^{PsyPAH} generates a stacking interaction with W196^{PsyPAH} (W188^{APF1}), closing the backside of the SBG, impeding the allocation of longer peptides. Residues L179^{PsyPAH} (Q171^{APF1}) and R183^{PsyPAH} (N175^{APF1}) would also hamper the presence of similar peptide ligands in PsyPAH (Figure 6). Finally, F204^{PsyPAH} (L196^{APF1}) and I207^{PsyPAH} (A199^{APF1}) reduce the SBG cleft volume, producing a higher hydrophobic character of this site compared to APF1, but lower than that presented by VinJ [18]. In fact, the hydrophobicity of this site would partly explain why PsyPAH can hydrolyze different aliphatic/aromatic amino acid amides, or even why the *p*-nitroanilide derivatives were hydrolyzed more efficiently than the amide derivatives (Table 2); the environment generated by Y186^{PsyPAH}, W196^{PsyPAH}, and F204^{PsyPAH} seems highly appropriate for the accommodation of an aromatic moiety. In this sense, it might be interesting to ascertain whereas other L-amino acid-amide derivatives with more voluminous amide substituents could be a more suitable starting material for their kinetic resolution using this subfamily of L-amidases. These differences in the SBG would support the different substrate specificity of PAHs when compared to PAPs.

3.5. Putative Catalytic Centre of PsyPAH

S113^{PsyPAH}, D253^{PsyPAH}, and H280^{PsyPAH} comprise the canonical clan SC class catalytic triad of the family (Table S3). The putative catalytic center of PsyPAH is buried into the structure, accessible through the deep hollow contiguous to the E1 site, where the substrate needs to enter to be cleaved. Whereas we were not able to obtain a ligand-bound structure through soaking experiments, an extra density was found at the S1 site in our crystallographic data, assigned as a phosphate molecule most likely arising from the initial purification buffer. This molecule is at a binding distance of N218^{PsyPAH} and E222^{PsyPAH} (Figures 7A and S5). Superposition with the APF1 bound to L-Proline reveals that both ligands occupy the same spatial position (Figure 7B). The counterpart of N209^{APF1} and E213^{APF1} residues were proved to be key for L-Pro-binding [55] together with Y205^{APF1} and E245^{APF1}. Since both enzymes process Pro-containing substrates and the four residues are conserved (N218^{PsyPAH}, E222^{PsyPAH}, Y214^{PsyPAH}, and E254^{PsyPAH}), a common L-Pro binding mode can be defined (Figure 7B).

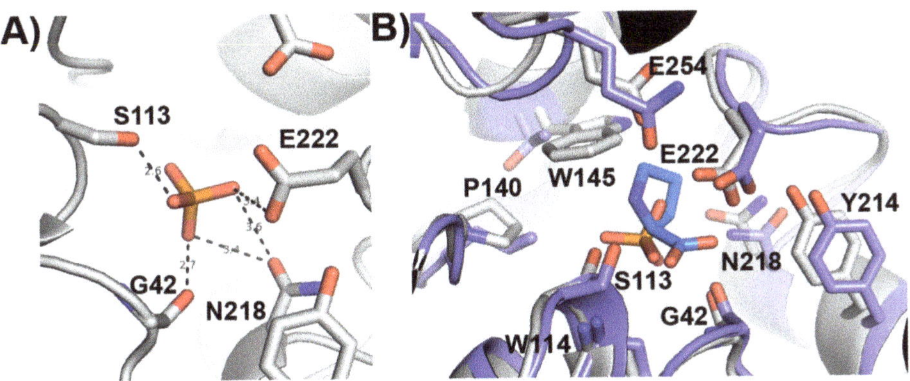

Figure 7. (**A**) Catalytic cleft of PsyPAH showing the modeled phosphate molecule. (**B**) Superposition of APF1 bound to L-Proline at the S1 site (PDB 1XRR, blue tones) and PsyPAH bound to phosphate (white tones). The numbering corresponds to PsyPAH residues.

Analogously to APF1, our structural model also supports the carbonyl/amide groups from the peptide bonds of G42^{PsyPAH} (G37^{APF1}) and W114^{PsyPAH} (Y106^{APF1}) as the constituents of the oxyanion hole (Figure 7A,B, Table S3). Despite the conservation of these key residues, the rest of the amino acids comprising S1 and S1′ sites are quite different (Table S3), while providing an overall hydrophobic character to these environments. It is important to highlight that P139^{PsyPAH} (L131^{APF1}) and W145^{PsyPAH} (T137^{APF1}) transform the PsyPAH S1 site into a much smaller and more hydrophobic cleft when compared to APF1, which might account for the substrate scope of PsyPAH toward different non-polar amides (see above).

Finally, further comparison of PsyPAH with SmPAP and XcPAP reveals that whereas the catalytic triad (S113^{PsyPAH}, D253^{PsyPAH}, and H280^{PsyPAH}) is positionally conserved in the catalytic domain, key binding residues of the cap domain (N218^{PsyPAH}, E222^{PsyPAH}, Y214^{PsyPAH}, and E254^{PsyPAH}, also present in APF1, VinJ, and MysPAP, Table S3), are not conserved. These results confirm the discussion about the PAP classification and provide additional clues on the different substrate scope observed among PAPs [56].

4. Conclusions

In conclusion, we report the first crystal structure of a PAP-like amidase (S33 peptidase clan) at 1.95 Å resolution with potential application within the "Amidase Process", showing a broad substrate specificity toward different canonical and non-canonical amino acids. Structural and sequence analyses allow one to decipher different L-amidase subfamilies, a prerequisite to finding enzymes with new or improved properties. Besides, the overall structure of PsyPAH is more similar to VinJ (a S33 peptidase, not a PAP), and structural comparison showed a higher conservation of key residues of the activity of APF1 (a S33 peptidase, not a strict PAP), suggesting a similar catalytic mechanism to that proposed for the latter. The lower volume and hydrophobicity of the S1 and E1 sites seem to account for the activity with smaller L-amino acid amides.

Therefore, our results confirm PsyPAH as a different member of the S33 peptidase family, which is not strictly a PAP enzyme. Future work should focus on understanding the substrate specificity of amidases conforming the S33 peptidase clan through mutational and structural studies. Since the divergence of the cap domain among these enzymes seems critical for substrate specificity, special attention should be taken to accurately classify them.

Supplementary Materials: The following are available online at https://www.mdpi.com/article/10.3390/cryst12010018/s1. Synthesis of different amino-acid amides. Figure S1: Kinetic determinations for L-Pro- and L-Leu-amide. Figure S2: Phylogenetic analysis of different enzymes with proven L-amidase activity. Figure S3: Surface representation of different peptidase S33 family members. Figure S4: Comparison of the polyketide substrate binding site of VinJ with that of APF1, MysPAP, and PsyPAH. Figure S5: Omit maps calculated for the PsyPAH structure. Table S1: Homolog structures of PsyPAH obtained with the DALI server. Table S2: Different ligand-bound structures of Tricorn Interacting Factor F1. Table S3: Residues involved in substrate binding and catalysis in the different pockets of APF1. Table S4: Comparison of residues proposed in the polyketide binding tunnel in VinJ.

Author Contributions: Conceptualization, S.M.-R. and J.A.G.; methodology, all authors; investigation, all authors; writing—original draft preparation, S.M.-R. and J.A.G.; writing—review and editing, all authors; funding acquisition, S.M.-R., R.C.-M., L.Á.d.C. and J.A.G. All authors have read and agreed to the published version of the manuscript.

Funding: This research was supported by the Spanish Ministry of Science and Innovation/FEDER funds grant PID2020-116261GB-I00/AEI/10.13039/501100011033 (JAG), from the FEDER/Junta de Andalucía-Consejería de Transformación Económica, Industria, Conocimiento y Universidades grants P18-FR-3533 (LAC) and P12-FQM-790 (RCM), and from the University of Granada grant PPJI2017-1 (SMR).

Institutional Review Board Statement: Not applicable.

Informed Consent Statement: Not applicable.

Data Availability Statement: Coordinates and structure factors have been deposited at the PDB with accession code 7A6G.

Acknowledgments: We are grateful to the European Synchrotron Radiation Facility (ESRF), Grenoble, France, for the provision of time through proposals Mx1938 and Mx2064, and the staff at ID30B beamline for their assistance during data collection. SMR and JTP are also grateful to the Andalusian Regional Government through the Endocrinology and Metabolism Group (CTS-202). We want to thank "Unidad de Excelencia Química aplicada a Biomedicina y Medioambiente" of the University of Granada.

Conflicts of Interest: The authors declare no conflict of interest. The funders had no role in the design of the study; in the collection, analyses, or interpretation of data; in the writing of the manuscript, or in the decision to publish the results.

References

1. Komeda, H.; Harada, H.; Washika, S.; Sakamoto, T.; Ueda, M.; Asano, Y. S-stereoselective piperazine-2-tert-butylcarboxamide hydrolase from *Pseudomonas azotoformans* IAM 1603 is a novel L-amino acid amidase. *Eur. J. Biochem.* **2004**, *271*, 1465–1475. [CrossRef] [PubMed]
2. Komeda, H.; Hariyama, N.; Asano, Y. L-Stereoselective amino acid amidase with broad substrate specificity from *Brevundimonas diminuta*: Characterization of a new member of the leucine aminopeptidase family. *Appl. Microbiol. Biotechnol.* **2006**, *70*, 412–421. [CrossRef] [PubMed]
3. Yamaguchi, S.; Komeda, H.; Asano, Y. New enzymatic method of chiral amino acid synthesis by dynamic kinetic resolution of amino acid amides: Use of stereoselective amino acid amidases in the presence of alpha-amino-epsilon-caprolactam racemase. *Appl. Environ. Microbiol.* **2007**, *73*, 5370–5373. [CrossRef] [PubMed]
4. Martínez-Rodríguez, S.; Torres, J.M.; Sánchez, P.; Ortega, E. Overview on Multienzymatic Cascades for the Production of Non-canonical α-Amino Acids. *Front. Bioeng. Biotechnol.* **2020**, *8*, 887. [CrossRef]
5. Hamer, D.; Greenstein, J.P. An enzymatic resolution of proline. *J. Biol. Chem.* **1951**, *193*, 81–89. [CrossRef]
6. Work, E.; Birnbaum, S.; Winitz, M.; Greenstein, J. Separation of the Three Isomeric Components of Synthetic α, ε-Diaminopimelic Acid1. *J. Am. Chem. Soc.* **1955**, *77*, 1916–1918. [CrossRef]
7. Sonke, T.; Ernste, S.; Tandler, R.F.; Kaptein, B.; Peeters, W.P.; van Assema, F.B.; Wubbolts, M.G.; Schoemaker, H.E. L-selective amidase with extremely broad substrate specificity from *Ochrobactrum anthropi* NCIMB 40321. *Appl. Environ. Microbiol.* **2005**, *71*, 7961–7973. [CrossRef] [PubMed]
8. van den Tweel, W.J.J.; van Dooren, T.J.G.M.; de Jonge, P.H.; Kaptein, B.; Duchateau, A.L.L.; Kamphuis, K. *Ochrobactrum anthropi* NCIMB 40321: A new biocatalyst with broad-spectrum l-specific amidase activity. *Appl. Microbiol. Biotechnol.* **1993**, *39*, 296–300. [CrossRef]

9. Hermes, H.F.; Sonke, T.; Peters, P.J.; van Balken, J.A.; Kamphuis, J.; Dijkhuizen, L.; Meijer, E.M. Purification and Characterization of an l-Aminopeptidase from *Pseudomonas putida* ATCC 12633. *Appl. Environ. Microbiol.* **1993**, *59*, 4330–4334. [CrossRef] [PubMed]
10. Hermes, H.F.; Tandler, R.F.; Sonke, T.; Dijkhuizen, L.; Meijer, E.M. Purification and Characterization of an l-Amino Amidase from *Mycobacterium neoaurum* ATCC 25795. *Appl. Environ. Microbiol.* **1994**, *60*, 153–159. [CrossRef]
11. Kamphuis, J.; Meijer, E.M.; Boesten, W.H.J.; Broxterman, Q.B.; Kaptein, B.; Hermes, H.F.M.; Schoemaker, H.E. Production of natural and synthetic L- and D-amino acids by aminopeptidases and amino amidases. In *Biocatalytic Production of Amino Acids and Derivatives*; Rozzell, J.D., Wagner, F., Eds.; Wiley: New York, NY, USA, 1992; pp. 178–206.
12. Maestracci, M.; Bui, K.; Thiéry, A.; Arnaud, A.; Galzy, P. The amidases from a *Brevibacterium strain*: Study and applications. *Adv. Biochem. Eng. Biotechnol.* **1988**, *36*, 67–115. [CrossRef] [PubMed]
13. Youshko, M.I.; Luuk, M.; Sheldon, R.A.; Švedas, V.K. Application of aminoacylase I to the enantioselective resolution of α-amino acid esters and amides. *Tetrahedron Asymmetry* **2004**, *15*, 1933–1936. [CrossRef]
14. Stelkes-Ritter, U.; Beckers, G.; Bommarius, A.; Drauz, K.; Günther, K.; Kottenhahn, M.; Schwarm, M.; Kula, M.R. Kinetics of Peptide Amidase and its Application for the Resolution of Racemates. *Biocatal. Biotransform.* **2009**, *15*, 205–219. [CrossRef]
15. Mahon, C.S.; O'Donoghue, A.J.; Goetz, D.H.; Murray, P.G.; Craik, C.S.; Tuohy, M.G. Characterization of a multimeric, eukaryotic prolyl aminopeptidase: An inducible and highly specific intracellular peptidase from the non-pathogenic fungus *Talaromyces emersonii*. *Microbiology* **2009**, *155*, 3673–3682. [CrossRef] [PubMed]
16. Inoue, T.; Ito, K.; Tozaka, T.; Hatakeyama, S.; Tanaka, N.; Nakamura, K.T.; Yoshimoto, T. Novel inhibitor for prolyl aminopeptidase from *Serratia marcescens* and studies on the mechanism of substrate recognition of the enzyme using the inhibitor. *Arch. Biochem. Biophys.* **2003**, *416*, 147–154. [CrossRef]
17. Tamura, T.; Tamura, N.; Lottspeich, F.; Baumeister, W. Tricorn protease (TRI) interacting factor 1 from *Thermoplasma acidophilum* is a proline iminopeptidase. *FEBS Lett.* **1996**, *398*, 101–105. [CrossRef]
18. Shinohara, Y.; Miyanaga, A.; Kudo, F.; Eguchi, T. The crystal structure of the amidohydrolase VinJ shows a unique hydrophobic tunnel for its interaction with polyketide substrates. *FEBS Lett.* **2014**, *588*, 995–1000. [CrossRef] [PubMed]
19. Ezawa, T.; Jung, S.; Kawashima, Y.; Noguchi, T.; Imai, N. Convenient green preparation of dipeptides using unprotected α-amino acids. *Tetrahedron Asymmetry* **2017**, *28*, 75–83. [CrossRef]
20. Riley, J.P. The spectrophotometric determination of ammonia in natural waters with particular reference to sea-water. *Anal. Chim. Acta* **1953**, *9*, 575–589. [CrossRef]
21. Erlanger, B.F.; Kokowsky, N.; Cohen, W. The preparation and properties of two new chromogenic substrates of trypsin. *Arch. Biochem. Biophys.* **1961**, *95*, 271–278. [CrossRef]
22. Kabsch, W. XDS. *Acta Cryst. D Biol. Cryst.* **2010**, *66*, 125–132. [CrossRef]
23. Winn, M.D.; Ballard, C.C.; Cowtan, K.D.; Dodson, E.J.; Emsley, P.; Evans, P.R.; Keegan, R.M.; Krissinel, E.B.; Leslie, A.G.; McCoy, A.; et al. Overview of the CCP4 suite and current developments. *Acta Cryst. D Biol. Cryst.* **2011**, *67*, 235–242. [CrossRef] [PubMed]
24. Bunkóczi, G.; Echols, N.; McCoy, A.J.; Oeffner, R.D.; Adams, P.D.; Read, R.J. Phaser.MRage: Automated molecular replacement. *Acta Cryst. D Biol. Cryst.* **2013**, *69*, 2276–2286. [CrossRef] [PubMed]
25. Afonine, P.V.; Grosse-Kunstleve, R.W.; Echols, N.; Headd, J.J.; Moriarty, N.W.; Mustyakimov, M.; Terwilliger, T.C.; Urzhumtsev, A.; Zwart, P.H.; Adams, P.D. Towards automated crystallographic structure refinement with phenix.refine. *Acta Cryst. D Biol. Cryst.* **2012**, *68*, 352–367. [CrossRef] [PubMed]
26. Murshudov, G.N.; Skubák, P.; Lebedev, A.A.; Pannu, N.S.; Steiner, R.A.; Nicholls, R.A.; Winn, M.D.; Long, F.; Vagin, A.A. REFMAC5 for the refinement of macromolecular crystal structures. *Acta Cryst. D Biol. Cryst.* **2011**, *67*, 355–367. [CrossRef]
27. Emsley, P.; Lohkamp, B.; Scott, W.G.; Cowtan, K. Features and development of Coot. *Acta Cryst. D Biol. Cryst.* **2010**, *66*, 486–501. [CrossRef] [PubMed]
28. Painter, J.; Merritt, E.A. Optimal description of a protein structure in terms of multiple groups undergoing TLS motion. *Acta Cryst. D Biol. Cryst.* **2006**, *62*, 439–450. [CrossRef]
29. Chen, V.B.; Arendall, W.B.; Headd, J.J.; Keedy, D.A.; Immormino, R.M.; Kapral, G.J.; Murray, L.W.; Richardson, J.S.; Richardson, D.C. MolProbity: All-atom structure validation for macromolecular crystallography. *Acta Cryst. D Biol. Cryst.* **2010**, *66*, 12–21. [CrossRef]
30. Laskowski, R.A.; Jabłońska, J.; Pravda, L.; Vařeková, R.S.; Thornton, J.M. PDBsum: Structural summaries of PDB entries. *Protein Sci.* **2018**, *27*, 129–134. [CrossRef]
31. Madeira, F.; Park, Y.M.; Lee, J.; Buso, N.; Gur, T.; Madhusoodanan, N.; Basutkar, P.; Tivey, A.R.N.; Potter, S.C.; Finn, R.D.; et al. The EMBL-EBI search and sequence analysis tools APIs in 2019. *Nucleic Acids Res.* **2019**, *47*, W636–W641. [CrossRef]
32. Robert, X.; Gouet, P. Deciphering key features in protein structures with the new ENDscript server. *Nucleic Acids Res.* **2014**, *42*, W320–W324. [CrossRef] [PubMed]
33. Letunic, I.; Bork, P. Interactive Tree Of Life (iTOL) v4: Recent updates and new developments. *Nucleic Acids Res.* **2019**, *47*, W256–W259. [CrossRef] [PubMed]
34. Holm, L. DALI and the persistence of protein shape. *Protein Sci.* **2020**, *29*, 128–140. [CrossRef] [PubMed]
35. The Pymol Molecular Graphics System, Schrödinger, LLC. Available online: https://www.scirp.org/(S(vtj3fa45qm1ean45vvffcz55))/reference/ReferencesPapers.aspx?ReferenceID=1958992 (accessed on 25 November 2021).
36. Stetefeld, J.; McKenna, S.A.; Patel, T.R. Dynamic light scattering: A practical guide and applications in biomedical sciences. *Biophys. Rev.* **2016**, *8*, 409–427. [CrossRef]

37. Wu, Z.; Liu, C.; Zhang, Z.; Zheng, R.; Zheng, Y. Amidase as a versatile tool in amide-bond cleavage: From molecular features to biotechnological applications. *Biotechnol. Adv.* **2020**, *43*, 107574. [CrossRef]
38. Wagner, F.W.; Wilkes, S.H.; Prescott, J.M. Specificity of *Aeromonas* aminopeptidase toward amino acid amides and dipeptides. *J. Biol. Chem.* **1972**, *247*, 1208–1210. [CrossRef]
39. Nandan, A.; Nampoothiri, K.M. Molecular advances in microbial aminopeptidases. *Bioresour. Technol.* **2017**, *245*, 1757–1765. [CrossRef]
40. Nandan, A.; Nampoothiri, K.M. Therapeutic and biotechnological applications of substrate specific microbial aminopeptidases. *Appl. Microbiol. Biotechnol.* **2020**, *104*, 5243–5257. [CrossRef]
41. Inoue, A.; Komeda, H.; Asano, Y. Asymmetric Synthesis of L-α-Methylcysteine with the Amidase from *Xanthobacter flavus* NR303. *Adv. Synth. Catal.* **2005**, *347*, 1132–1138. [CrossRef]
42. Matsushima, M.; Takahashi, T.; Ichinose, M.; Miki, K.; Kurokawa, K.; Takahashi, K. Structural and immunological evidence for the identity of prolyl aminopeptidase with leucyl aminopeptidase. *Biochem. Biophys. Res. Commun.* **1991**, *178*, 1459–1464. [CrossRef]
43. Nakamura, T.; Yu, F. Amidase Gene. U.S. Patent 6617139, 9 September 2003.
44. Katoh, O.; Akiyama, T.; Nakamura, T. Novel Amide Hydrolase Gene. EP Patent Application EP1428876A1, 16 June 2004.
45. Asano, Y. L-Amino Acid Amide Asymmetric Hydrolase and Dna Encoding the Same. EP1770166A4, 19 December 2007.
46. Matsushima, M.; Takahashi, T.; Ichinose, M.; Miki, K.; Kurokawa, K.; Takahashi, K. Prolyl aminopeptidases from pig intestinal mucosa and human liver: Purification, characterization and possible identity with leucyl aminopeptidase. *Biomed. Res.* **1991**, *12*, 323–333. [CrossRef]
47. Rawlings, N.D.; Barrett, A.J.; Thomas, P.D.; Huang, X.; Bateman, A.; Finn, R.D. The MEROPS database of proteolytic enzymes, their substrates and inhibitors in 2017 and a comparison with peptidases in the PANTHER database. *Nucleic Acids Res.* **2018**, *46*, D624–D632. [CrossRef]
48. Holmquist, M. Alpha/Beta-hydrolase fold enzymes: Structures, functions and mechanisms. *Curr. Protein Pept. Sci.* **2000**, *1*, 209–235. [CrossRef] [PubMed]
49. Marchot, P.; Chatonnet, A. Enzymatic activity and protein interactions in alpha/beta hydrolase fold proteins: Moonlighting versus promiscuity. *Protein Pept. Lett.* **2012**, *19*, 132–143. [CrossRef] [PubMed]
50. Goettig, P.; Groll, M.; Kim, J.S.; Huber, R.; Brandstetter, H. Structures of the tricorn-interacting aminopeptidase F1 with different ligands explain its catalytic mechanism. *EMBO J.* **2002**, *21*, 5343–5352. [CrossRef]
51. Lenfant, N.; Hotelier, T.; Velluet, E.; Bourne, Y.; Marchot, P.; Chatonnet, A. ESTHER, the database of the α/β-hydrolase fold superfamily of proteins: Tools to explore diversity of functions. *Nucleic Acids Res.* **2013**, *41*, D423–D429. [CrossRef]
52. Medrano, F.J.; Alonso, J.; García, J.L.; Romero, A.; Bode, W.; Gomis-Rüth, F.X. Structure of proline iminopeptidase from *Xanthomonas campestris* pv. citri: A prototype for the prolyl oligopeptidase family. *EMBO J.* **1998**, *17*, 1–9. [CrossRef] [PubMed]
53. Yoshimoto, T.; Kabashima, T.; Uchikawa, K.; Inoue, T.; Tanaka, N.; Nakamura, K.T.; Tsuru, M.; Ito, K. Crystal structure of prolyl aminopeptidase from *Serratia marcescens*. *J. Biochem.* **1999**, *126*, 559–565. [CrossRef]
54. Okai, M.; Miyauchi, Y.; Ebihara, A.; Lee, W.C.; Nagata, K.; Tanokura, M. Crystal structure of the proline iminopeptidase-related protein TTHA1809 from *Thermus thermophilus* HB8. *Proteins* **2008**, *70*, 1646–1649. [CrossRef] [PubMed]
55. Goettig, P.; Brandstetter, H.; Groll, M.; Göhring, W.; Konarev, P.V.; Svergun, D.I.; Huber, R.; Kim, J.S. X-ray snapshots of peptide processing in mutants of tricorn-interacting factor F1 from *Thermoplasma acidophilum*. *J. Biol. Chem.* **2005**, *280*, 33387–33396. [CrossRef]
56. Ito, K.; Inoue, T.; Kabashima, T.; Kanada, N.; Huang, H.S.; Ma, X.; Azmi, N.; Azab, E.; Yoshimoto, T. Substrate recognition mechanism of prolyl aminopeptidase from *Serratia marcescens*. *J. Biochem.* **2000**, *128*, 673–678. [CrossRef] [PubMed]

Article

Shifts in Backbone Conformation of Acetylcholinesterases upon Binding of Covalent Inhibitors, Reversible Ligands and Substrates

Zoran Radić

Skaggs School of Pharmacy and Pharmaceutical Sciences, University of California San Diego, La Jolla, CA 92093-0751, USA; zradic@ucsd.edu

Abstract: The influence of ligand binding to human, mouse and *Torpedo californica* acetylcholinesterase (EC 3.1.1.7; AChE) backbone structures is analyzed in a pairwise fashion by comparison with X-ray structures of unliganded AChEs. Both complexes with reversible ligands (substrates and inhibitors) as well as covalently interacting ligands leading to the formation of covalent AChE conjugates of tetrahedral and of trigonal-planar geometries are considered. The acyl pocket loop (AP loop) in the AChE backbone is recognized as the conformationally most adaptive, but not necessarily sterically exclusive, structural element. Conformational changes of the centrally located AP loop coincide with shifts in C-terminal α-helical positions, revealing interacting components for a potential allosteric interaction within the AChE backbone. The stabilizing power of the aromatic choline binding site, with the potential to attract and pull fitting entities covalently tethered to the active Ser, is recognized. Consequently, the pull can promote catalytic reactions or relieve steric pressure within the impacted space of the AChE active center gorge. These dynamic properties of the AChE backbone inferred from the analysis of static X-ray structures contribute towards a better understanding of the molecular template important in the structure-based design of therapeutically active molecules, including AChE inhibitors as well as reactivators of conjugated, inactive AChE.

Keywords: acetylcholinesterase; organophosphate; carbamate; backbone conformation; oxime reactivation; oxime antidote; acetylcholine

Citation: Radić, Z. Shifts in Backbone Conformation of Acetylcholinesterases upon Binding of Covalent Inhibitors, Reversible Ligands and Substrates. *Crystals* **2021**, *11*, 1557. https://doi.org/10.3390/cryst11121557

Academic Editor: Abel Moreno

Received: 1 December 2021
Accepted: 11 December 2021
Published: 14 December 2021

1. Introduction

Acetylcholinesterase (EC 3.1.1.7; AChE) is an essential regulatory enzyme in cholinergic neurotransmission of vertebrates. It has evolved, consistent with its important function, into near-perfect biological catalyst [1]. The high catalytic throughput in the hydrolysis of the neurotransmitter acetylcholine (ACh) is achieved owing to the specific three-dimensional structure of its catalytic subunit. Whether completion of the catalytic cycle in AChE depends on conformational adjustments in its structure is an open question.

Because of the size of the ~70 kDa catalytic subunit, only approaches capable of resolving "static" structural snapshots have been successful in obtaining atomic structures of AChE [2]. Those structural snapshots, documented in more than 200 PDB-deposited X-ray structures, have not revealed any significant conformational outliers and seem to largely reveal one dominant conformation of the backbone [3]. Several smaller conformational deviations have been identified, however. Most of them have been associated with the AChE acyl pocket loop [4,5]. Because macromolecular structural dynamics and the catalytic activity of AChE can be detected even in the crystalline state [6,7], analysis of low-level conformational diversity in AChE X-ray structures could provide a glimpse of a larger-scale flexibility requirement for physiological function in solution.

Catalytic specificities of cholinesterases, both AChE and that of the closely related butyrylcholinesterase (EC 3.1.1.8; BChE), have been previously defined to depend on the size and ligand-binding capacity of three enzyme domains: the acyl pocket, the choline binding site and the peripheral site [8]. We have recently revealed, however, that the size of the acyl pocket in human AChE (hAChE) has to be used with caution as a sole criterion for the specificity [9]. We have shown that some of large organophosphate (OP) substrates of hAChE, such as Novichok A234, contrary to expectation, insert their ethoxy substituent on phosphorus into the hAChE acyl pocket and fit the overall active center without notable distortions of the hAChE backbone [9] or its side-chains.

In here, the presented analysis of complexes and conjugates of AChEs is expanded to include those with carboxylic and carbamic ester substrates in addition to OPs. Furthermore, in addition to the structure of hAChE, the structures of mouse AChE (mAChE) and Torpedo californica AChE (TcAChE) have been analyzed. In a pairwise fashion, the PDB-deposited X-ray structures of native AChEs have been compared to those of their carbamylated conjugates, OP-hAChE conjugates as well as of those reversibly complexed with carboxylic acid substrates, hydrolytic products and some of the tight binding reversible inhibitors. The Pairwise Alpha Carbon Comparison Tool (PACCT 3) was used for comparisons (www.ZENODO.org, doi:10.5281/zenodo.3992329. Last access on 30 November 2021), as presented earlier [9–12].

2. Materials and Methods

X-ray structures of hAChE, mAChE and TcAChE in their unliganded, apo forms, covalently inhibited by OPs or carbamates, in reversible complexes with the substrates ACh, acetylthiocholine (ATCh), butyrylthiocholine (BTCh), hydrolytic products choline (Ch), thiocholine (TCh) and acetate (Ac) and with tight binding reversible inhibitors, were obtained from the RCSB PDB database [13]. The frames of reference in pairwise structural comparisons were the structures of unliganded (apo) AChEs; this was, specifically, PDB ID 4EY4 for hAChE [14], PDB ID 1J06 for mAChE [6] and PDB ID 2ACE for TcAChE [2].

The Pairwise Alpha Carbon Comparison Tool (PACCT 3) used for pairwise comparisons of α-carbon positions/shifts in structural backbones is available for download from Zenodo (www.ZENODO.org, doi:10.5281/zenodo.3992329. Last access on 30 November 2021), and was used as shown earlier [9–12]. A bar chart PACCT 3 output of shifts in backbone Cα positions (given in Å) was used for comparison as a function of amino acid residue number. For each of the analyzed structures, unless otherwise indicated, chain A was compared to the chain A of the respective apo-AChE. The resulting bar chart was overlaid with charts of other relevant structures in question. Average Cα shift values were indicated in bar charts for each of the inter-structural comparisons as horizontal dashed lines. Peaks of Cα shifts extending above the inter-structural average were taken as significant. The exceptions were peaks observed at the C- and N-termini and around breaks (missing parts of compared structures).

For the visual analysis of the structures in this comparison, PACCT 3-generated overlays were used and visualized in the VR-based visualization suite Nanome (Nanome Inc., San Diego, CA, USA). Overlays were based on the superposition of twenty Cα atom coordinates with the smallest observed values of Cα shifts in the respective comparison.

3. Results

3.1. *Effects of Covalent Binding on Backbone Conformation of hAChE*

3.1.1. Reversible, Covalent Binding of the Substrate Analogue 4KTMA

Pairwise analyses of covalent conjugates of the hemiketal 4KTMA (Figure 1A) with hAChE, mAChE and TcAChE, in comparison to structures of their corresponding apo forms, do not reveal any significant backbone distortions (Figure 1). Only small ~0.2 Å Cα shifts could be observed in the more N-terminal part of the Ω loop and in the C-terminally located α-helix II in all 3 AChEs (Figure 1B). The bound 4KTMA entity fills and fits well in the lower part of the active center gorge between the acyl pocket and the choline binding site. Its geometry mimics the tetrahedral transition state in the hydrolytic cycle of ACh. The similarity of conjugated backbone conformations with those of apo AChEs suggest that active center geometries of native, apo hAChEs appear quite adequate for achieving fast catalytic turnovers of the bound physiological substrate without additional substrate-induced adjustments.

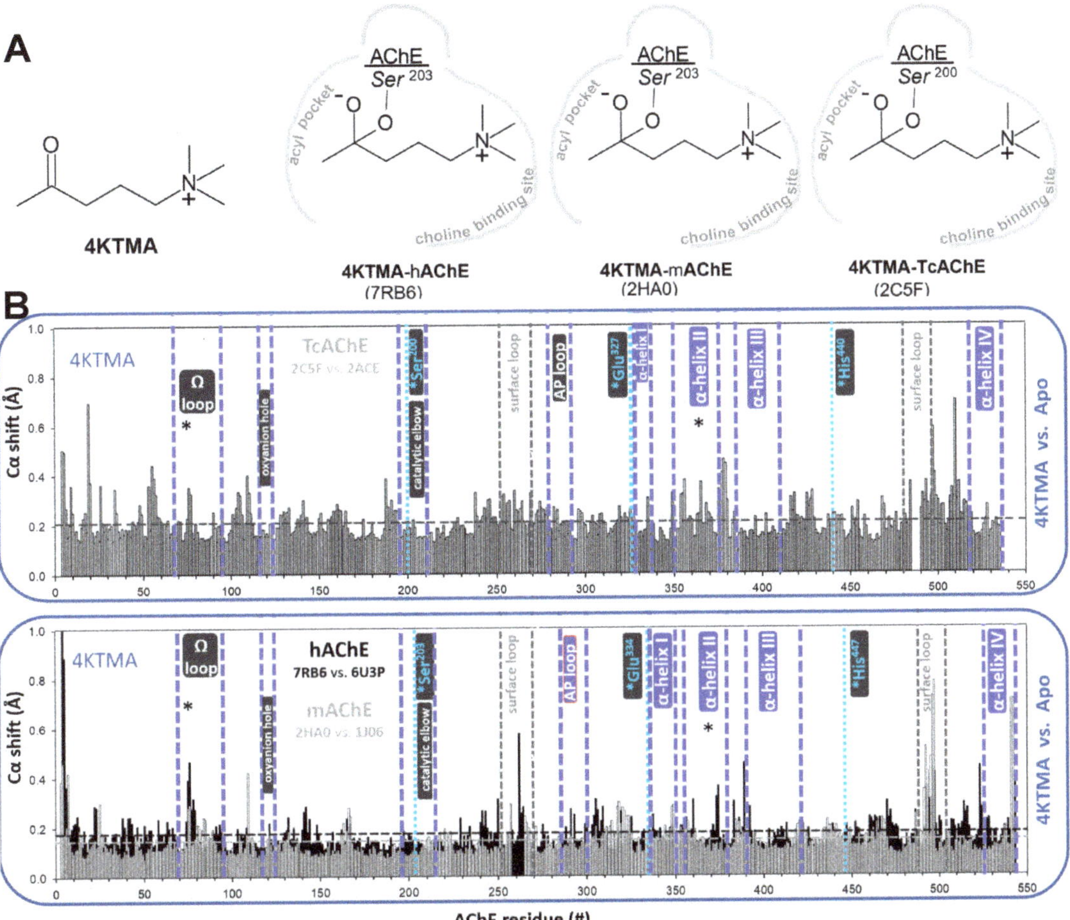

Figure 1. (**A**) Structure of the substrate analogue 4KTMA and schematic representation of its binding to the active serine within the active centers of three AChEs based on X-ray structures with PDB codes given in parentheses. (**B**) Bar charts from PACCT3 analyses of each of the three conjugate structures compared to the corresponding apo AChE forms. Alpha carbon shifts (in Å) of conjugated backbones are represented as vertical bars for each amino acid in the linear AchE sequence. Horizontal dashed lines indicate an average backbone atom shift, while stars indicate shifts considered as significant. Functionally important structural elements of AChE are bracketed between vertical, dashed, and blue lines and labelled. Their location within the 3D structure of AChE is shown in Figure 2.

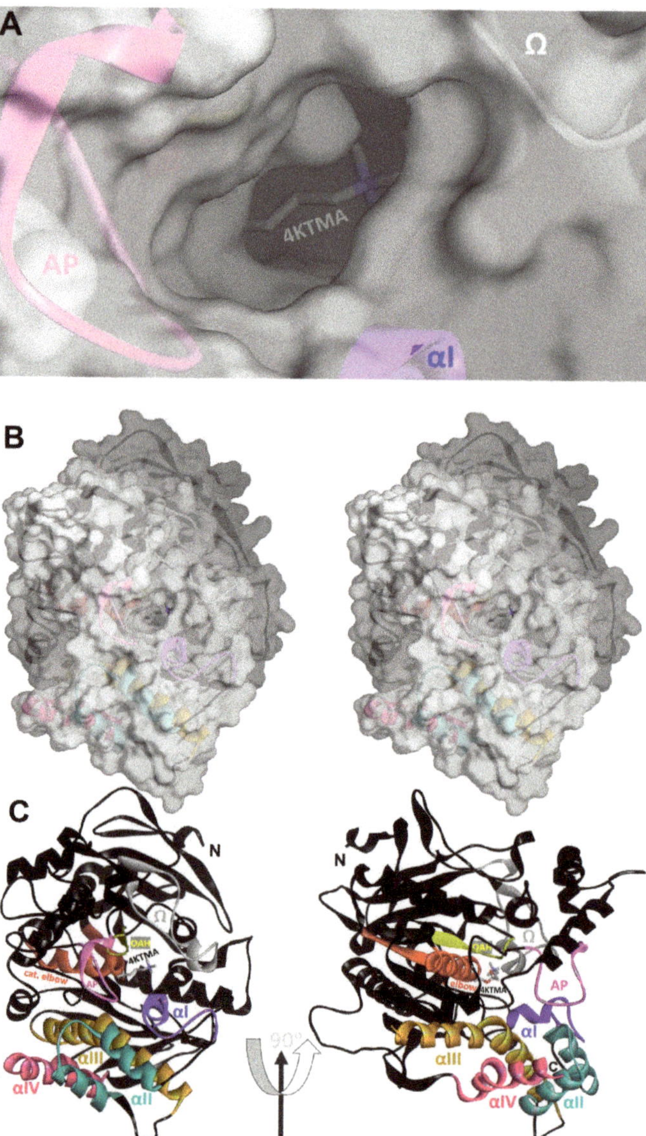

Figure 2. Locations of structural elements of the hAChE backbone affected by covalent or reversible ligand binding as detected by PACCT 3 analysis. (**A**) Close-up of the active center gorge opening in the 4KTMA-hAChE conjugate (PDB ID 7RB6). Semi-transparent Connolly surface presentation. (**B**) Stereo view of the catalytic subunit. The semi-transparent surface reveals Cα ribbon color coded by structural elements and in the same view as in the (**C**) **Left panel:** showing AP loop (286–299; pink), Ω-loop (69–96; gray), oxyanion hole (117–124; yellow), catalytic elbow (196–215; red), α-helices I (336–350; blue), II (356–382; teal), III (390–421) orange), and IV (525–543; magenta). (**C**) **Right panel:** the same color-coded elements in ribbon, rotated clockwise by 90°. Covalently bound 4KTMA is rendered as sticks.

In particular, conformations of three loops that control the size of the active center opening (AP loop, Ω loop and α-helix I; cf. Figure 2) do not seem to be influenced in the tetrahedral 4KTMA conjugate (Figure 1B) in any concerted manner.

3.1.2. Covalent Conjugates Formed by OPs

Tetrahedral conjugates formed by covalent binding of OPs have the capacity to significantly influence backbone conformations of AChEs, as already noted for hAChE [9]. One of two ethoxy groups of diethylphosphorylated hAChE, formed by POX inhibition, has to be accommodated in the small acyl pocket of AChE resulting in a large distortion of the AP loop (Figure 3). This was observed in both of two different POX-hAChE X-ray structure datasets (5HF5 and 5HF8), but also in the structure of the smaller, dealkylated (aged) POX-hAChE conjugate (5HF6). The two-point attachment of those conjugates, one by covalent bond to Ser203 and two by tight stabilization of the phosphonyl oxygen in the oxyanion hole, should, however, still allow for a shift of a conjugate towards the empty space of the choline binding site in order to relieve the steric AP loop pressure, but that is not observed in these X-ray structures. Additionally, no backbone shifts are observed around two attachment points, the oxyanion hole and Ser203 (Figure 3), indicating no tendency for the conjugate to shift in spite of the availability of the space within the active center gorge for that shift. It seems that the distorted AP loop is not forcing the offending ethoxy group out of the acyl pocket in order to resume its native conformation. This possibly reflects a small energetic difference between a variety of conformations of this loop and speaks of the freedom of its intrinsic conformational flexibility.

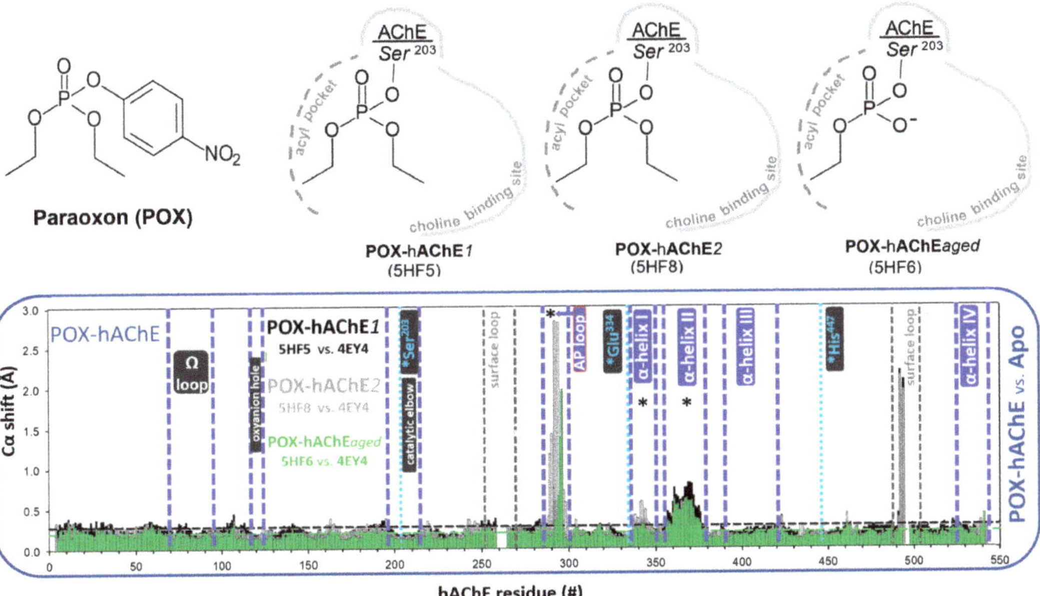

Figure 3. Structure of paraoxon (POX), an active form of the pesticide parathion, and schematic representation of its covalent binding to the active serine of hAChE in two conformationally different covalent conjugates and in the aged (dealkylated) form. Schematics are based on X-ray structures with PDB codes given in parentheses. Bar charts from PACCT3 analyses of each of 3 structures compared to the apo AChE (4EY4). Alpha carbon shifts (in Å) of conjugated backbones are represented as vertical bars for each amino acid in the linear AChE sequence. Horizontal dashed lines indicate an average backbone atom shift, while stars indicate shifts considered as significant. Functionally important structural elements of hAChE, bracketed between vertical, dashed, and blue lines and labelled, are mapped within the 3D structure of the hAChE in Figure 2.

We recently suggested [9] that substituents on phosphorus capable of forming stabilizing interactions with the choline binding site (hydrophobic or electrostatic) can pull the respective conjugate away from the acyl pocket and allow the AP loop to recover its native conformation. In fact, a similar effect can be observed even when the acyl pocket-binding substituent is small and fits inside the AP pocket well, such as in the absence of steric clashes within the AP (Figure 4). For example, the binding of neither the methyl phosphonates, VX, soman or A230 distorts the AP loop (Figure 4). The P atom in those conjugates is, however, consistently pulled towards the choline binding site by virtue of hydrophobic and electrostatic attractions of the large choline binding site interacting with substituents in soman and A230. Phosphorus atoms in the conjugates of large soman and A230 are shifted by 0.5–0.6 Å compared to the much smaller VX.

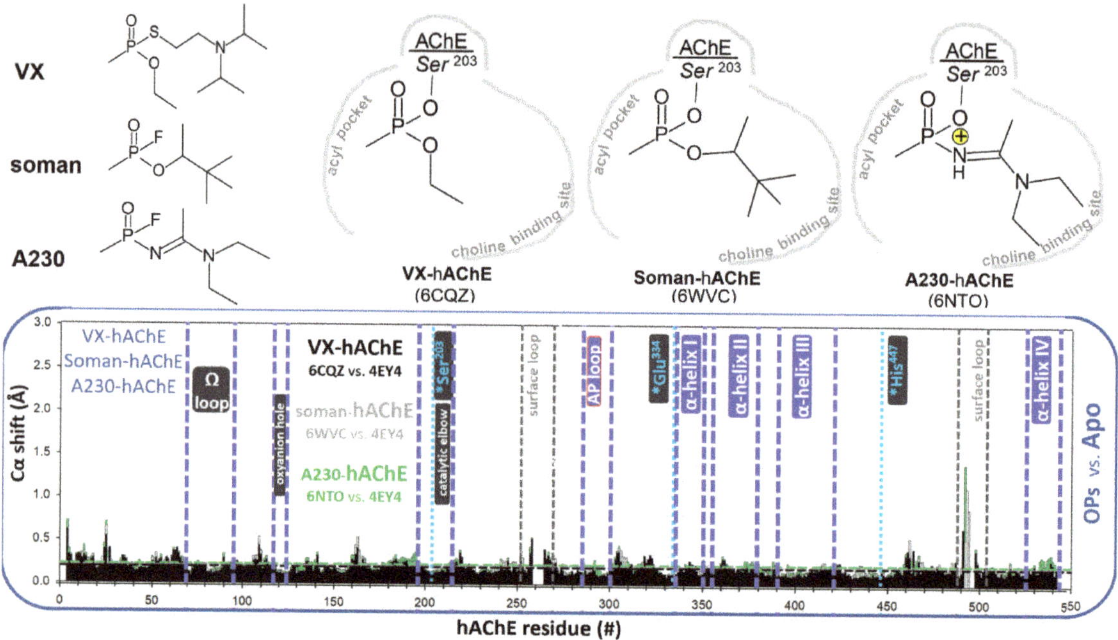

Figure 4. The structure of the nerve agents OPs, VX, soman and A230 and a schematic representation of their covalent binding to the active serine of hAChE. Schematics are based on X-ray structures with PDB codes given in parentheses. Bar charts from PACCT3 analyses of each of the three structures compared to the apo AChE (4EY4). Alpha carbon shifts (in Å) of conjugated backbones are represented as vertical bars for each amino acid in the linear AChE sequence. Horizontal dashed lines indicate an average backbone atom shift, while stars indicate shifts considered significant. Functionally important structural elements of hAChE, bracketed between vertical, dashed, and blue lines and labelled, are mapped within the 3D structure of hAChE in Figure 2.

The AP loop distortions seem to clearly coincide with shifts of the α-helix II, and to a smaller extent, the α-helix I (Figure 3). In the absence of the AP loop distortion, α-helices I and II do not shift (Figure 4). Those helices are within 5–6 Å of the AP loop, allowing for a direct interaction leading to conformational connectivity. This kind of connectivity between the active center and C-terminal α-helices of the hAChE homodimer interface is likely responsible for the POX-triggered dissociation of the hAChE homodimer, observed in solution using SAXS [11], indicating connectivity between the AP loop distortion and disturbance at the homodimer interface of hAChE.

In this context, it is also instructive to analyze differences in backbone conformations of OP-hAChE conjugates obtained by the inhibition of separate enantiomers of VX, S_PVX-hAChE (6CQZ) and R_PVX-hAChE (6CQX), since inhibition of hAChE by an R_P isomer of sarin (unlike the S_P isomer) was shown to have a similar effect on the homodimer dissociation as POX [11]. The enantiomeric S_PVX-hAChE conjugate was obtained by inhibition of racemic VX, due to the about two orders of magnitude faster rate of inhibition by the S_P enantiomer [15]. Indeed, consistent with AP loop distortions observed in the POX-hAChE conjugate (Figure 3), the R_PVX-hAChE structure showed both AP loop distortion and shifts in α-helices II and I, while none of those effects were observed in the structure of the S_PVX-hAChE conjugate (Figure 5). The enantio-specific effect is very clear in spite of some differences in the AP loop conformations in chains A and B of the R_PVX-hAChE (Figure 5), which in fact reflects its intrinsic conformational flexibility.

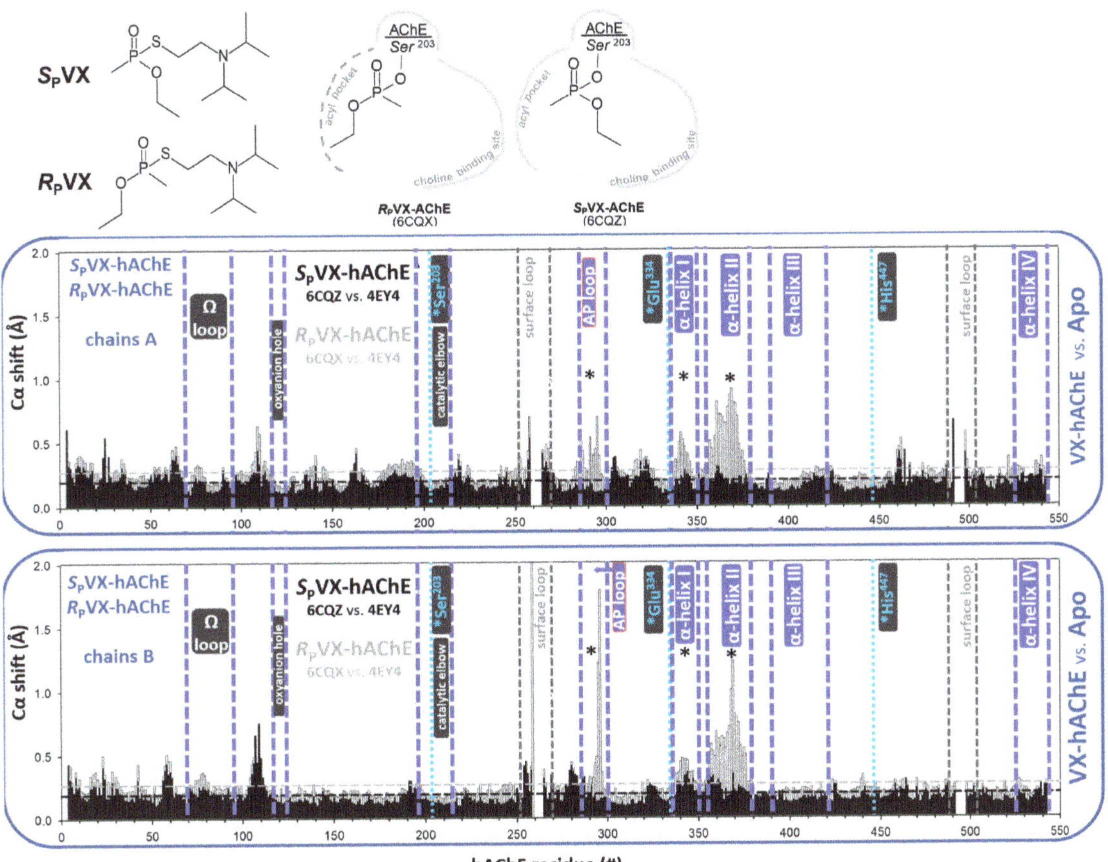

Figure 5. Structure of enantiomers of VX, and schematic representation of their covalent binding to the active serine of hAChE. Schematics are based on X-ray structures with PDB codes given in parentheses. Bar charts from PACCT3 analyses of each of three structures compared to the apo AChE (4EY4). Alpha carbon shifts (in Å) of conjugated backbones are represented as vertical bars for each amino acid in the linear AChE sequence. Horizontal dashed lines indicate an average backbone atom shift, while stars indicate shifts considered significant. Functionally important structural elements of hAChE, bracketed between vertical, dashed, and blue lines and labelled, are mapped within the 3D structure of hAChE in Figure 2.

Another commonly used symmetric OP, DFP (Figure 6), forms conjugates with AchEs where the AP loop is distorted upon the insertion of one of its two large isopropoxy substituents on phosphorus. It has been well described that DFP-AChEs age (dealkylate) in minutes to hours [16]. Available X-ray structures of both non-aged and aged DFP-AChE conjugates with mAChE and aged DFP-TcAChE conjugate reveal large distortions of their AP loops (Figure 6). Magnitudes of backbone Cα shifts appear similar to the ones inflicted by POX. The two different X-ray datasets of the aged DFP-mAChE (2JGM and 5HCU) differ somewhat in their extent of distortion, yet distortion is observed in both, along with associated shifts of α helices II and I (Figure 6). Furthermore, in one of the structures of the aged conjugate, the dealkylated group was shown to be stabilized within the AP (5HCU). This is somewhat surprising, considering the accepted mechanism of aging, where amino acid residues involved in the stabilization of carbonium cations in the course of dealkylation are the choline binding site residues F338 and E202 [16]. Interestingly, reversible binding of non-oxime reactivating antidote SP-134 stabilized one very distorted AP loop conformation upon its insertion into an already distorted acyl pocket of the aged DFP-mAChE (Figure 6). This effect is opposite from what is observed upon binding of pyridinium oxime reactivators HI-6 and/or 2PAM to POX-hAChE, where reversal of the distorted AP loop conformations into native conformations was shown [9]. The likely reason for the difference is that pyridinium oximes bind within the main cavity of the active center gorge of POX-hAChE without entering into the acyl pocket.

Conjugation of an AChE with tabun can also lead to distortion of the AP loop, but under slightly more complicated conditions. This was demonstrated recently in X-ray structures of TcAChE conjugated by NEDPA, an analog of tabun that forms identical conjugates with TcAChE [5]. The dimethylamino group in those conjugates (6G17 for Tabun or 6G4O; chain B for NEDPA conjugates) is always accommodated in the acyl pocket of the undisturbed, native conformation, while the larger ethoxy group points towards the choline binding site. Only reversible binding of novel uncharged aldoximes R1 or R2 to the conjugate was shown to have the capacity to distort the AP loop conformation. The structures of R1 and R2 are very similar [5]. Both reactivators were shown to distort only one of two monomers in the crystallographic homodimer of released structures. For R2, binding to chain A did not cause distortion, though it did for the chain B. On the other hand, R1 was found bound to only chain A where the AP loop was distorted, and chain B was free from both R1 binding and distortion (Figure 7). Those observations illustrate small magnitudes of energy barriers in the conformational flexibility of the AP loop, where reversibly bound R2 can, but does not have to, promote loop distortion. Both R1 and R2 bind to the same area of the peripheral site in TcAChE and in very similar orientations, where they interact directly with the concave parts of the AP loop. Furthermore, a significant observation related to the connectivity of the structural elements of TcAChE is that distortion of the AP loop always coincides with backbone shifts of α-helix II. When the AP loop is in its native conformation (no peaks in the bar chart in the AP loop area), no shifts (no peaks in bar charts) can be observed for α-helix II, as well (Figure 6). This seems to be a common connectivity pattern for both hAChE, mAChE and TcAChE.

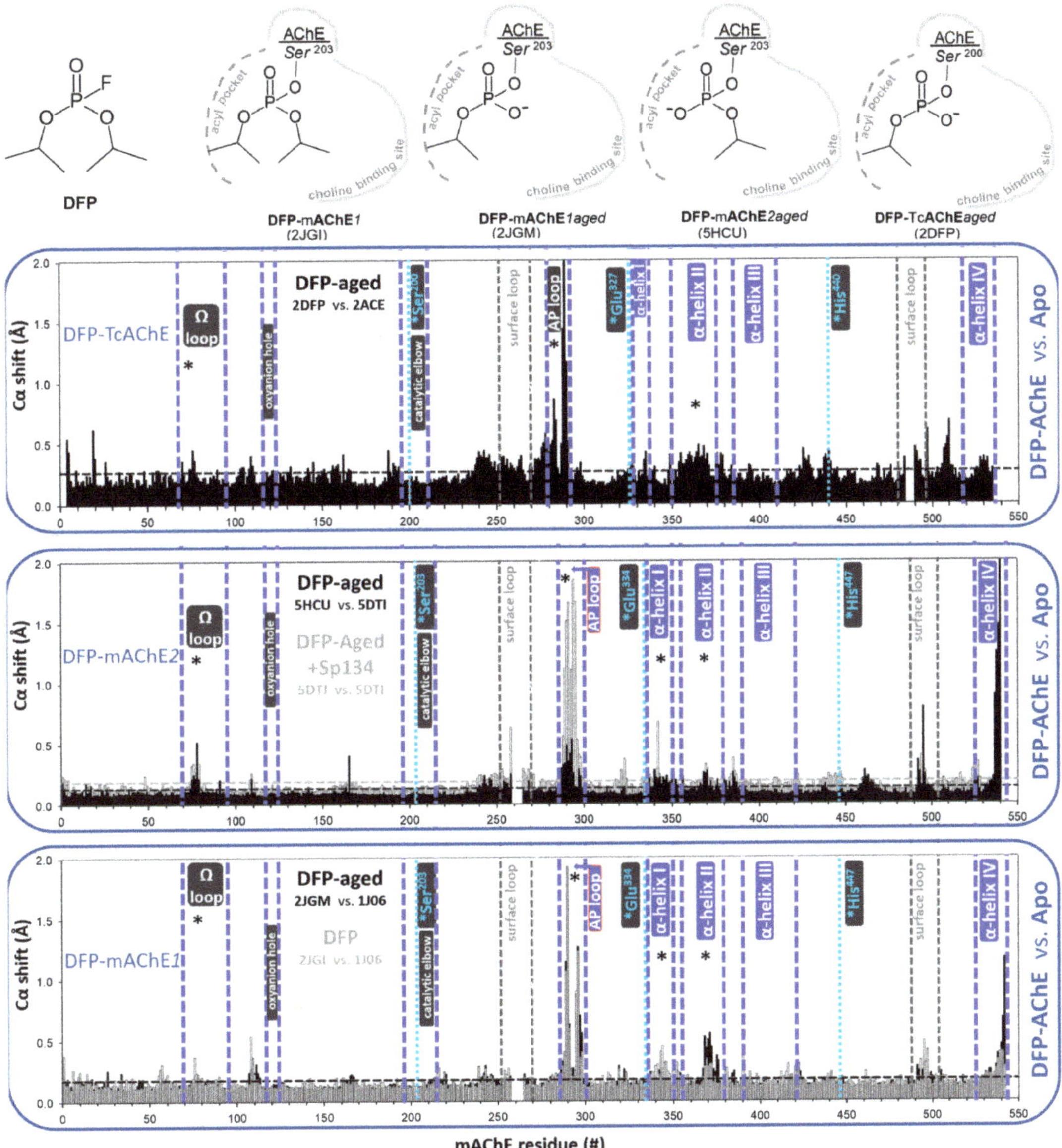

Figure 6. Structure of DFP and schematic representation of its covalent binding to the active serine of TCAChE and mAChE in aged (dealkylated) and non-aged forms. Schematics are based on one aged DFP-TcAChE and two sets of X-ray structures of DFP-mAChE with PDB codes given in parentheses. One of these structures refers to the reversible complex of aged DFP-mAChE with reactivator Sp134. Bar charts from PACCT3 analyses of each of these three structures compared to the corresponding apo AChEs. Alpha carbon shifts (in Å) of conjugated backbones are represented as vertical bars for each amino acid in the linear AChE sequence. Horizontal dashed lines indicate an average backbone atom shift, while stars indicate shifts considered significant. Functionally important structural elements of hAChE, bracketed between vertical, dashed, and blue lines and labelled, are mapped in the 3D structure of hAChE in Figure 2.

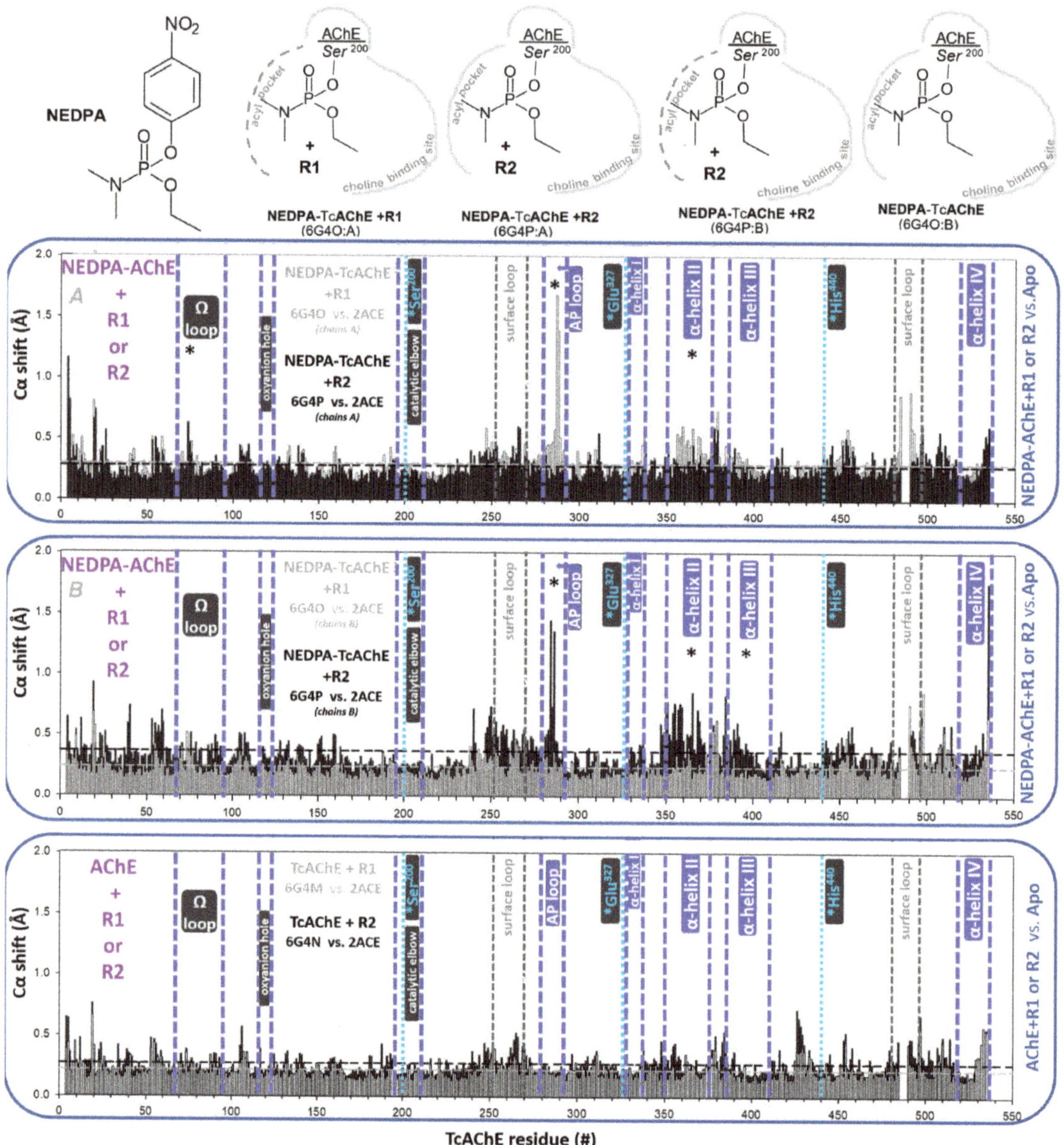

Figure 7. Structure of tabun analog NEDPA and schematic representation of its covalent binding to the active serine of TCAChE in the absence and the presence of oxime reactivators R1 and R2 [5]. Conjugates of NEDPA with AChE are identical to those formed by tabun. Schematics are based on X-ray structures of NEDPA-TcAChE conjugates, with PDB codes given in parentheses. Each structure contains chains A and B of the TcAChE homodimer in the asymmetric subunit. Monomers (chains) in homodimers show different conformations and were therefore analyzed by PACCT3 separately, with resulting data displayed in two separate bar charts. In bar charts, the alpha carbon shifts (in Å) of conjugated backbones are represented as vertical bars for each amino acid in the linear AChE sequence. Horizontal dashed lines indicate an average backbone atom shift, while stars indicate shifts considered significant. Functionally important structural elements of TcAChE are bracketed between vertical, dashed, and blue lines and labelled.

3.1.3. Covalent Conjugates Formed by Carbamates

The structures of carbamylated AChEs have been primarily studied with TcAChE. A largely planar configuration around covalently attached carbonyl carbon in carbamyl-AChE conjugates, compared to a tetrahedral configuration around covalently attached phosphorus in OP conjugates, seems to have an effect on the ability of substituents to be accommodated in the AP before the AP loop becomes distorted. For example, ethyl methyl-carbamylated TcAChE shows only small distortions of the AP loop backbone (Figure 8), compared to the much larger effect of the POX-inhibited hAChE (Figure 3) with similarly sized substituents (ethoxy group vs. ethyl methyl carbamyl group) accommodated in the pocket. The solvent accessible volume of the acyl pocket, at the same time, changes more significantly in the carbamylated TcAChE (Figure 9, panel C vs. panel A) and is similar to the effect in POX-hAChE (Figure 9, panel B vs. panel D) due to side-chain rotations. Nevertheless, the conformational flexibility of the AP loop backbone in carbamylated TcAChE remains large. Reversible binding of the above-mentioned uncharged reactivators R1 and R2 [5] to the peripheral site in TcAChE stabilizes the AP loop in two largely distorted conformations (Figure 7). Coincidentally, the positions of α-helices II are also clearly shifted, an effect already observed for OP-AChE conjugates. In addition, the most C-terminal α-helix IV shows notable shifts for the ethyl methyl carbamylated TcAChE (Figure 8).

The covalent binding of large monocarbamates, ganstigmine and MF268 to the active Ser200 of TcAChE appears to have no consequences on backbone conformations of TcAChE (Figure 8). Due to the larger rotational freedom in bonds of monocarbamates, they seem to achieve sterically fitting conformations inside the A easier than dicarbamates. In addition, the long aliphatic chain of MF268 seems to have sufficient intrinsic flexibility to avoid the steric obstacles of the narrow TcAChE active center gorge.

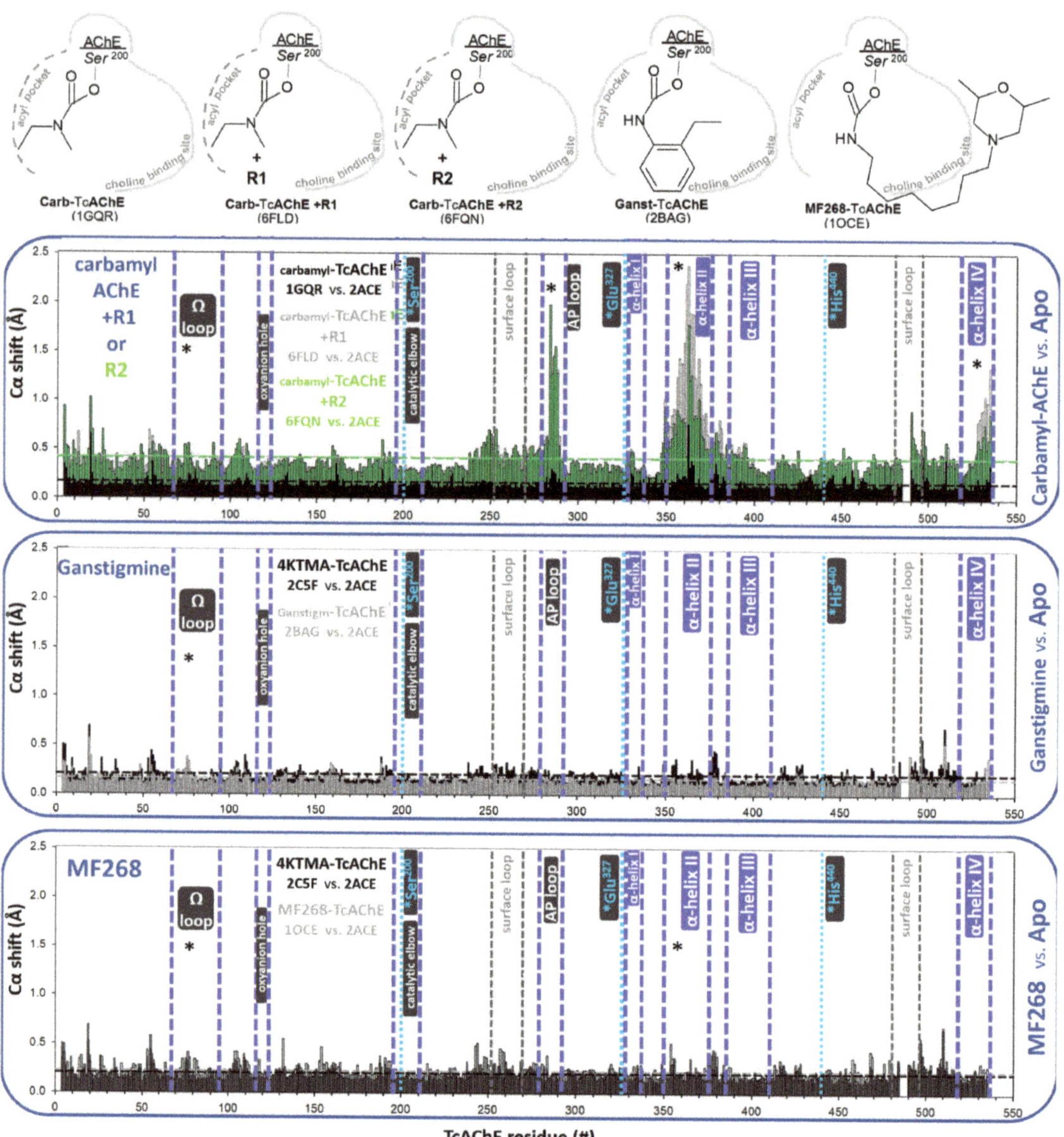

Figure 8. Schematic representation of TcAChE carbamylated by methyl ethylcarbamate, ganstigmine and carbamate MF268. Methylethylcarbamyl-TCAChE is shown in the presence of oxime reactivators R1 and R2 [5]. Schematics are based on X-ray structures with PDB codes given in parentheses. In bar charts, the alpha carbon shifts (in Å) of conjugated backbones are represented as vertical bars for each amino acid in the linear AChE sequence. Horizontal dashed lines indicate an average backbone atom shift, while stars indicate shifts considered significant. Functionally important structural elements of TcAChE are bracketed between vertical, dashed, and blue lines and labelled.

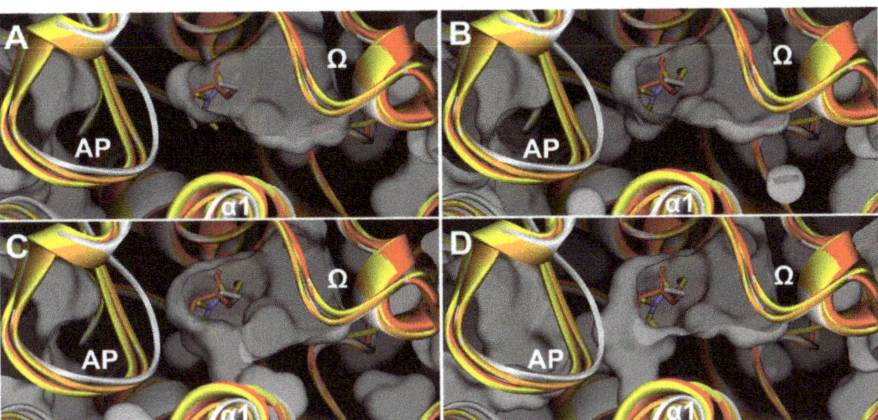

Figure 9. Comparative effects of ethyl methyl carbamylation of TcAChE and diethyl phosphorylation of hAChE on conformations of respective AP loops. Overlay of carbamyl-TcAChE (1GQR; yellow ribbon), apo-TcAChE (2ACE; red ribbon), POX-hAChE (5HF5; white ribbon), and apo-hAChE (4EY4; orange ribbon). Conjugated ethyl methylcarbamyl (yellow C atoms) and diethyl phosphoryl (white C atoms) entities are rendered as sticks. Shown are the Connolly surfaces of (**A**) Apo-TcAChE, (**B**) Apo-hAChE, (**C**) carbamyl-TcAChE and (**D**) POX-hAChE.

3.2. Effects of Reversible Binding on Backbone Conformation of hAChE

3.2.1. Reversible Binding of Substrates and Products

Structures of substrates of ACh, its thio analog ATCh and hydrolytic products Ch^-, TCh^- and Ac^- have been studied in complex with both TcAChE and mAChE [6,7]. The larger substrates BTCh and succinylcholine were studied only in complexes with mAChE [6]. While only a native, catalytically active enzyme was used in TcAChE studies, both a native and catalytically inactive Ser203Ala (S203A) mutant of mAChE was used. Crystallographic experiments contained all ligands in tens of mM to hundreds of mM concentrations, sufficient to allow them to form complexes in both the active center and in the peripheral site of AChEs, located at the rim of the active center gorge opening and adjacent to the Ω-loop. In all of those X-ray structures, the peripheral site was, therefore, occupied with a ligand, reflected in small shifts in the part of the Ω-loop (Figures 10 and 11). No other significant changes in Cα positions were observed except for shifts in α-helix II in the structure of the BTCh*mAChE complex, which were likely caused by an additional BTCh molecule bound from the outside to the surface of the enzyme molecule. This is reminiscent of the 4KTMA-AChE structures (Figure 1), except that in those tetrahedral conjugates, α-helix II was always slightly shifted. Both observations are consistent with the conclusion that conformations of AChEs observed in their apo states reflect conformations that are optimal for catalytic activity. Therefore, no conformational adjustments of the AChE backbone are needed when binding substrates to achieve the outstanding catalytic throughput of AChE.

On the other hand, the position of the covalently attached carbonyl carbon seems to have changed by ~0.8 Å from the acetylated form of TcAChE captured in 2HA5 to the tetrahedral form in conjugate with the transition state analogue 4KTMA (2HA0) in the direction of the choline binding site (Figure 12). Although slightly smaller in magnitude, this shift is analogous to the one observed for conjugated phosphorus in large OP-hAChE conjugates (Figure 4, [9]). It seems, therefore, that one of important properties of the choline binding site in AChEs is to attract and pull (and stabilize) large ester substituents, in both covalent inhibition and in the rapid catalytic cycle of an AChE.

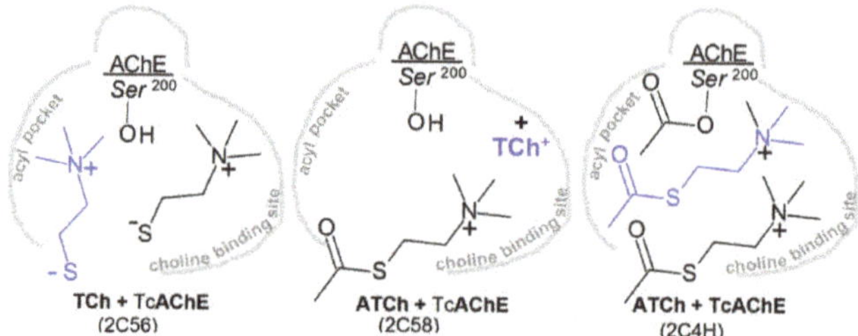

Figure 10. Structure of substrates ACh and ATCh and product TCh in reversible complexes with TcAChE. Schematics are based on X-ray structures, with PDB codes given in parentheses. Bar charts from PACCT3 analyses of each of three structures compared to the apo AChE (2ACE) are given in Figure S1.

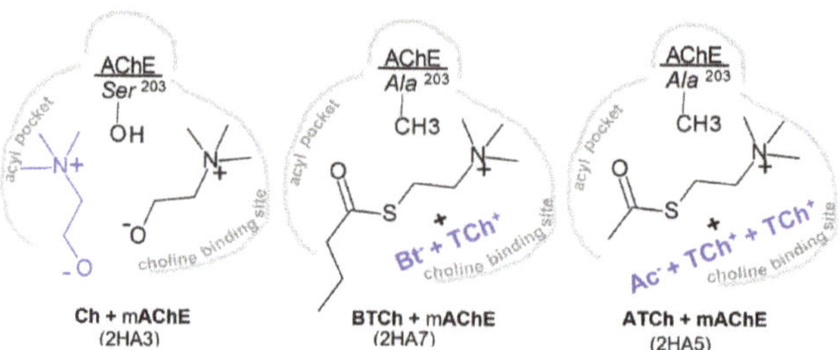

Figure 11. Structure of substrates ATCh and BTCh and product Ch in reversible complexes with mAChE. Schematics are based on X-ray structures with PDB codes given in parentheses. Bar charts from PACCT3 analyses of each of three structures compared to apo AChE (1J06) are given in Figure S2.

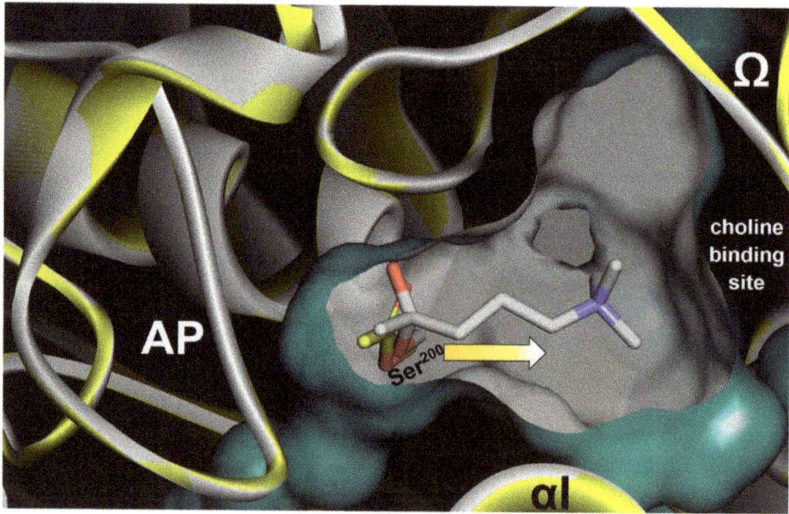

Figure 12. Shifts in position of the carbonyl carbon of the acetylated Ser^{200} of TcAChE (from 2C4H; yellow C atom sticks and yellow ribbon) towards the choline binding site (as indicated by the arrow) in the tetrahedral intermediate (4KTMA-TcAChE conjugate; white C atom sticks and white ribbon; 2C5F) by ~0.8 Å. Shown is the Connolly surface of the 2C4H cut-away to expose the active center gorge interior around active Ser200. Locations of the acyl pocket (AP), choline binding site, Ω-loop and α-helix I (αI) are indicated.

3.2.2. Reversible Binding of High-Affinity Inhibitors

While reversibly bound substrates, their hydrolytic products or transition state analogs, with high µM to mM binding affinity, do not induce notable changes in the backbone conformations of TcAChE and mAChE, a higher-affinity reversible ligand could have that capacity. The tightest binding of reversible ligands of AChEs are click-chemistry triazoles [17], created in situ within the AChE active center gorge and serving as a reaction vessel. Their binding affinities are in the fM concentration range. Due to the syn- vs. anti-isomerism of the triazole ring, a single pair of acetylene- and azide-decorated reactants can form either one of two isomers in situ (Figure 13). The highest affinity one is syn-TZ2PA6 (Figure 13), and TcAChE and mAChE, probably due to their intrinsically curved shape of the active center gorge channel, preferentially form syn-triazoles. There is a notable difference in the backbone configurations of structures of the syn- compared to anti-isomers (Figure 13). While anti-isomers practically do not affect the backbone conformation of mAChE, several orders of magnitude tighter-binding syn- isomers affect Ω-loop conformation and the conformations of α-helices II and I. Most of the Ω-loop distortions can also be observed in complexes with starting materials for the click-reaction (TZ2 and PA5; Figure 13). However, α-helix shifts are more pronounced in already formed syn-triazoles (both TZ2PA5 and TZ2PA6; Figure 13). It seems, therefore, that anti-triazoles bind best to the apo conformation of mAChE, and higher affinity syn-triazoles prefer mAChE of a slightly different backbone conformation.

Unlike high affinity triazoles that stabilize their tetrahydroacridine domains within the choline binding site and extend the phenantridinium ring system up the active center gorge towards its opening and the Ω-loop, picomolar inhibitor Tacrine-benzofuran extends its bi-cycle downwards, into the acyl pocket, near the base of the TcAChE active center gorge [18]. As a consequence, the conformation of the AP loop is severely distorted in this complex, with a significant shift of the α-helix II (Figure 14). This is likely due to a tight interaction between the benzofuran moiety with the aromatic residues of the AP pocket base, of the nature previously studied in other biological systems [19,20]. Backbone conformations of the Ω-loop or other relevant structural elements are not affected. The magnitude of the AP loop shift is larger than what was previously reported for bis-tacrine5 [21], another tacrine-based reversible inhibitor (Figure S3). While the choline binding site stabilizes the tetrahydroacridine moiety similarly in both compounds, the second tetrahydroacridine of bis-tacrine5 interacts with the concave part of the AP loop in a way that is similar to the interaction of R2 (Figure 8). Both R2 and bis-tacrine5 stabilize a similarly distorted conformation of the AP loop. In addition, the effect of bis-tacrine5 on the α-helix II shift is even larger than the one observed for tacrine-benzofuran (Figures 14 and S3). Taken together, this is another example where shifts in the α-helix II are inflicted indirectly via an AP loop interaction, since neither ligand makes direct contact with the helix. The assumption of connectivity of the two structural elements (the AP loop and C-terminal α-helix II) in the direction of the lower part of the active center gorge towards the enzyme's C-terminus (where the homo-dimerization interface is located), thus holds for reversible ligands, as well.

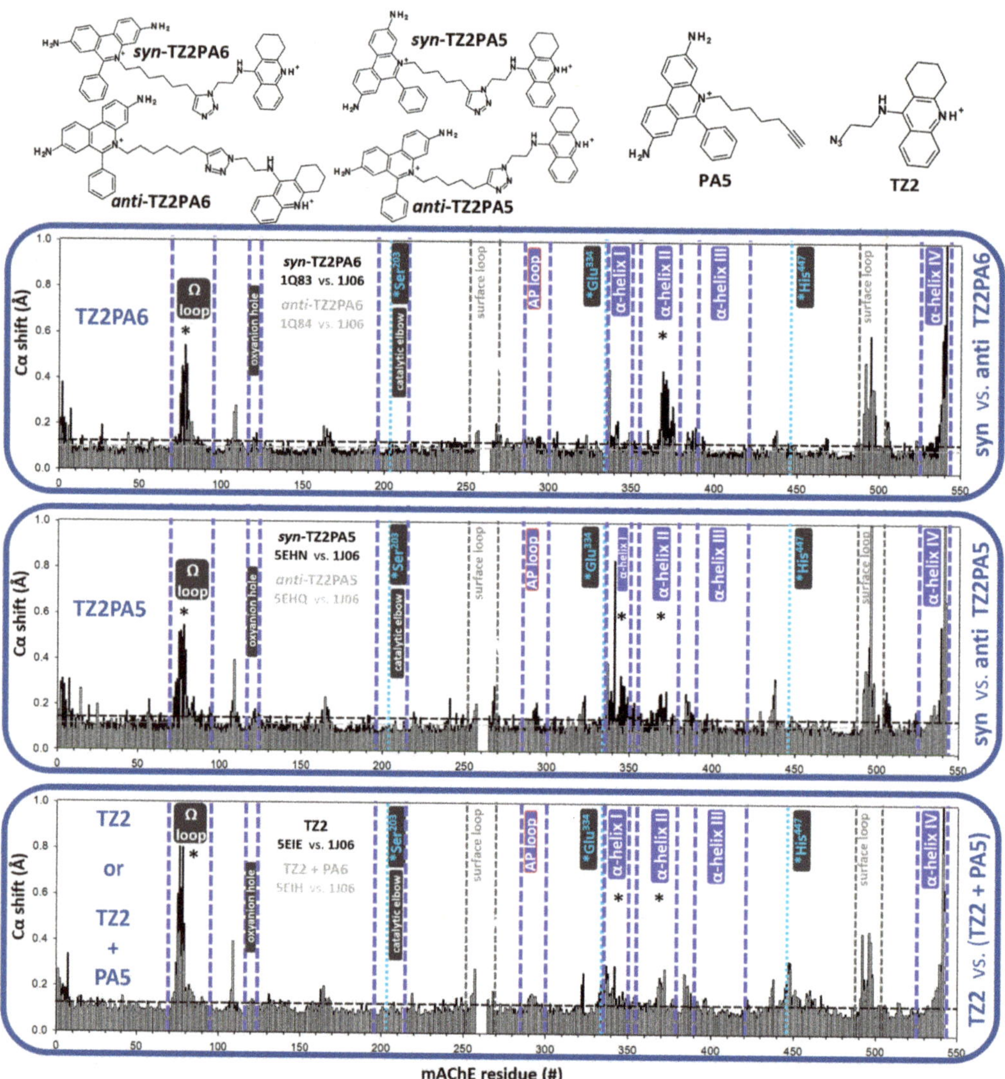

Figure 13. Structures of tight-binding, reversible, triazole inhibitors of AChE. Syn- and anti-isomers differ by several orders of magnitude in their binding affinities. TZ2 and PA6 are the starting materials for the in situ click-chemistry synthesis of triazoles. Compounds are shown in protonation states predominant at the physiological pH 7.4. Bar charts are from PACCT3 analyses of each of six structures compared to the apo AChE (1J06). Alpha carbon shifts (in Å) of conjugated backbones are represented as vertical bars for each amino acid in the linear AchE sequence. Horizontal dashed lines indicate an average backbone atom shift, while stars indicate shifts considered significant. Functionally important structural elements of hAChE, bracketed between vertical, dashed, blue lines and labelled, are mapped within the 3D structure of hAChE in Figure 2.

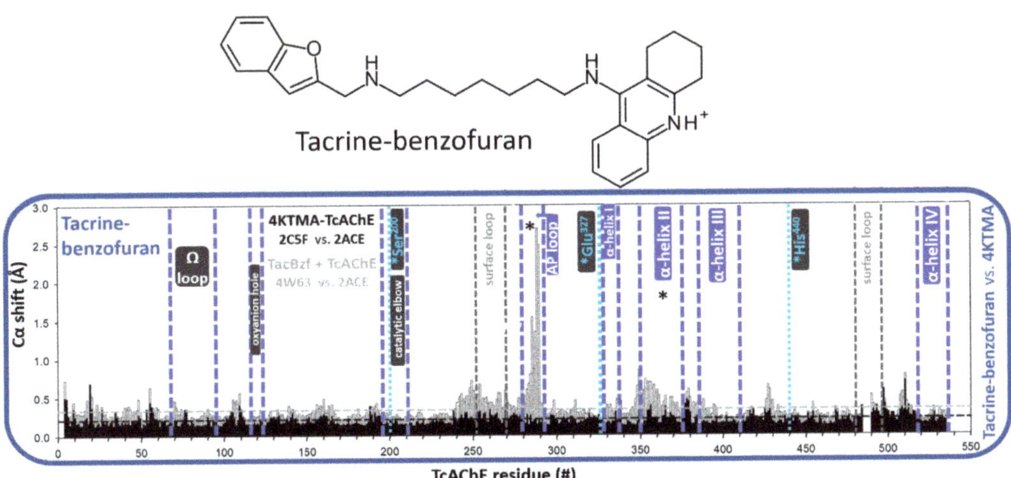

Figure 14. Structure of tacrine-benzofuran, a pM reversible inhibitor of TcAChE. Bar charts from PACCT3 analyses of the reversible complex X-ray structure compared to the apo AChE (2ACE). Parallel comparison of 4KTMA-TcAChE conjugate vs. apo TcAChE (2ACE) is given as a frame of reference. The compound is shown in protonation states predominant at the physiological pH 7.4. Alpha carbon shifts (in Å) of conjugated backbones are represented as vertical bars for each amino acid in the linear AChE sequence. Horizontal dashed lines indicate an average backbone atom shift, while stars indicate shifts considered significant. Functionally important structural elements of hAChE are bracketed between vertical, dashed, and blue lines and labelled.

4. Discussion

The present study presented pairwise analyses of changes in the Cα backbone conformations of non-liganded hAChE, mAChE and TcAChE upon their interaction with reversible and covalent ligands, providing evidence for several function-related properties of AChE's structure.

In interactions with physiological and closely related substrates (Ach, ATCh, BTCh), and, in particular, with the tetrahedral transition state analog 4KTMA, no significant alterations of the Cα backbone conformations were observed, suggesting that conformations of the apo forms of AchEs determined by X-ray diffraction generally reflect an optimal geometry evolved for its highly advanced catalytic mechanism. This is also consistent with our previous room-temperature X-ray crystallography of the 4KTMA-hAChE conjugate, which revealed a basically unchanged conformation of the hAChE backbone in diffraction data collected at either 100 K or 293 K [10].

The effects of a larger diversity of tetrahedral conjugate geometries on backbone conformations in AChEs are revealed through the investigation of OP esters that dephosphorylate very slowly and are thus amenable for X-ray structure determination. In principle, whenever a substituent larger than two carbons (the equivalent of a propionyl substituent in propionylcholine) is directed into the AP from the P atom tethered to the active Ser by a covalent bond and by phosphonyl oxygen stabilization in the oxyanion hole, the conformation of the AP loop is distorted. This is demonstrated for POX-hAChE and DFP-AChEs in both aged and non-aged conjugates, as well as in tabun-AChEs (NEDPA-AChE) and R_PVX-hAChE. A notable exception is A234-hAChE, a Novichok conjugate where AP loop distortion is avoided by the 1.0–1.5 Å shift of the conjugated OP towards the choline binding site. A property of the latter site to attract and pull suitable entities tethered to the active Ser is revealed even for OP conjugates with small AP-directed substituents, such as VX, soman and A230 (Figure 3). The presence of phosphorus in soman and A230 causes a shift in those conjugates of 0.5–0.6 Å compared to VX. Due to the larger size of their ester substituents directed towards the choline binding site, they have a better potential for tighter hydrophobic stabilization within that site. A similar 0.8 Å shift of the Ser-conjugated carbonyl carbon can be seen in a comparison of the acetyl-TcAChE and 4KTMA-TcAChE conjugates (Figure 12), where the much larger "choline-like" quaternary ammonium entity of 4KTMA fits snugly into the TcAChE choline binding site by virtue of a cation-Π interaction. The choline binding site thus seems to have the function of not only directing, but also of pulling and "stretching-out" Ser-conjugated entities for more optimal catalysis. On the other hand, the conformation of the AP loop does not seem to be rigid to the extent of steric expulsion of large substituents. It is, for example, flexible enough to distort its conformation in the aged POX-hAChE or DFP-AChEs, with no or with minimal shifts of conjugated phosphorus in the direction of the choline binding site. Dealkylation-generated sizeable empty spaces within the choline binding site thus remain unoccupied, while on the other side, the AP loop is distorted, with a seemingly small energy cost. While the binding of reversible ligands, such as oximes 2PAM or HI6, can revert a distorted AP loop into its native conformation, certain ligands that can either fit into the occupied AP or bind within the concave part of the AP loop have the capacity to stabilize and further potentiate distortion, as observed for SP134 and reactivators R1 and R2 bound to tabun (or NEDPA)-TcAChE conjugates.

Shifts of C-terminal α-helices are always observed alongside distortion of the AP loop, revealing an important connectivity that could lead to a component of an allosteric interaction. Those shifts, however, are typically not reverted upon reversible ligand binding.

The planar geometry around the atom conjugated to Ser, found in acetylated-TcAChE but also in carbamyl-TcAChE conjugates, results in similar AP loop distortions for large substituents, such as for ethyl methyl carbamate resulting from inhibition by rivastigmine. The same mentioned reactivators, R1 and R2, further potentiate this distortion, as in tabun-TcAChE conjugates. Connectivity with C-terminal α-helices is again observed. The large mono-substituted carbamates, ganstigmine and MF268, however, direct their substituents away from the AP loop, and thus, do not inflict distortion and do not affect the shift in the C-terminal α-helices.

Finally, into the emerging picture of a flexible AP loop, connected C-terminal α-helices and an attracting choline binding site, the non-covalent complexation of tight binding inhibitors fits well. Complexes of different triazole isomers prefer slightly different conformations of mAChE in the Ω-loop and C-terminal α-helices, while tacrine-benzofuran in direct contact with the AP loop stabilizes its distorted conformation and indirectly induces C-terminal α-helical shifts.

In conclusion, the analysis of backbone conformations in X-ray structures of three AChEs (hAChE, mAChE and TcAChE) interacting with diverse ligands reveals the AP loop backbone conformation as the most responsive and conformationally diverse of all structural elements of the AChE backbone. Structural changes in this loop located in the immediate vicinity of the catalytic triad Ser practically always coincide with shifts of one or more C-terminally located α-helices with the capacity to establish connectivity that could result in an allosteric interaction. AP loop conformations do not appear to be rigid and sterically exclusive, but rather adjustable around both small and large interacting ligands. The stabilizing potential of the choline binding site, on the other side of the active center gorge, on the other hand, acts as an attractant for large hydrophobic and cationic entities that can pull them out of the spatially restricted AP and relieve overall compaction within the active center gorge of AChEs. It is reasonable to assume that, in solution, the magnitudes of conformational freedom and the described interactions increase, representing significant constraints in the structure-based design of both therapeutical inhibitors and reactivators of already inhibited AChEs.

Supplementary Materials: The following are available online at https://www.mdpi.com/article/10.3390/cryst11121557/s1, Figure S1: Analysis of substrates ACh and ATCh and product TCh in reversible complexes with TcAChE. Figure S2: Analysis of substrates ATCh and BTCh and product Ch in reversible complexes with mAChE, Figure S3: Analysis of Bis-Tacrine5 reversibly bound to TcAChE.

Funding: This research was supported by the CounterACT Program, National Institutes of Health Office of the Director (NIH OD), the National Institute of Neurological Disorders and Stroke (NINDS), [Grant Numbers U01 NS083451 and R21 NS098998] and by the UCSD Academic Senate grant BG084144.

Data Availability Statement: All data are contained within the manuscript or are available to be shared upon request (please contact the author).

Conflicts of Interest: The author declares no conflict of interest.

References

1. Quinn, D.M. Acetylcholinesterase: Enzyme structure, reaction dynamics, and virtual transition states. *Chem. Rev.* **1987**, *87*, 955–979. [CrossRef]
2. Sussman, J.L.; Harel, M.; Frolow, F.; Oefner, C.; Goldman, A.; Toker, L.; Silman, I. Atomic structure of acetylcholinesterase from Torpedo californica: A prototypic acetylcholine-binding protein. *Science* **1991**, *253*, 872–879. [CrossRef] [PubMed]
3. Radić, Z.; Taylor, P. Structure and Function of Cholinesterases. In *Toxicology of Organophosphate and Carbamate Compounds*; Gupta, R.C., Ed.; Elsevier: San Diego, CA, USA, 2006.
4. Franklin, M.C.; Rudolph, M.J.; Ginter, C.; Cassidy, M.S.; Cheung, J. Structures of paraoxon-inhibited human acetylcholinesterase reveal perturbations of the acyl loop and the dimer interface. *Proteins* **2016**, *84*, 1246–1256. [CrossRef] [PubMed]
5. Santoni, G.; de Sousa, J.; de la Mora, E.; Dias, J.; Jean, L.; Sussman, J.L.; Silman, I.; Renard, P.Y.; Brown, R.; Weik, M.; et al. Structure-Based Optimization of Nonquaternary Reactivators of Acetylcholinesterase Inhibited by Organophosphorus Nerve Agents. *J. Med. Chem.* **2018**, *61*, 7630–7639. [CrossRef] [PubMed]
6. Bourne, Y.; Radić, Z.; Sulzenbacher, G.; Kim, E.; Taylor, P.; Marchot, P. Substrate and product trafficking through the active center gorge of acetylcholinesterase analyzed by crystallography and equilibrium binding. *J. Biol. Chem.* **2006**, *281*, 29256–29267. [CrossRef] [PubMed]
7. Colletier, J.P.; Fournier, D.; Greenblatt, H.M.; Stojan, J.; Sussman, J.L.; Zaccai, G.; Silman, I.; Weik, M. Structural insights into substrate traffic and inhibition in acetylcholinesterase. *EMBO J.* **2006**, *25*, 2746–2756. [CrossRef] [PubMed]
8. Radić, Z.; Pickering, N.A.; Vellom, D.C.; Camp, S.; Taylor, P. Three distinct domains in the cholinesterase molecule confer selectivity for acetyl- and butyrylcholinesterase inhibitors. *Biochemistry* **1993**, *32*, 12074–12084. [CrossRef] [PubMed]
9. Luedtke, S.; Bojo, C.; Li, Y.; Luna, E.; Pomar, B.; Radić, Z. Backbone Conformation Shifts in X-ray Structures of Human Acetylcholinesterase upon Covalent Organophosphate Inhibition. *Crystals* **2021**, *11*, 1270. [CrossRef]
10. Gerlits, O.; Blakeley, M.P.; Keen, D.A.; Radić, Z.; Kovalevsky, A. Room temperature crystallography of human acetylcholinesterase bound to a substrate analogue 4K-TMA: Towards a neutron structure. *Curr. Res. Struct. Biol.* **2021**, *3*, 206–215. [CrossRef] [PubMed]
11. Blumenthal, D.K.; Cheng, X.; Fajer, M.; Ho, K.Y.; Rohrer, J.; Gerlits, O.; Taylor, P.; Juneja, P.; Kovalevsky, A.; Radić, Z. Covalent inhibition of hAChE by organophosphates causes homodimer dissociation through long-range allosteric effects. *J. Biol. Chem.* **2021**, *297*, 1–14. [CrossRef]
12. Rohrer, J.; Sidhom, M.; Han, J.; Radić, Z. Overlay-independent comparisons of X-ray structures reveal small, systematic conformational changes in liganded acetylcholinesterase. *Period. Biol.* **2016**, *118*, 319–328. [CrossRef]
13. wwPDB Consortium. Protein Data Bank: The single global archive for 3D macromolecular structure data. *Nucleic Acids Res.* **2019**, *47*, D520–D528. Available online: https://pubmed.ncbi.nlm.nih.gov/30357364/ (accessed on 10 December 2021). [CrossRef] [PubMed]
14. Cheung, J.; Rudolph, M.J.; Burshteyn, F.; Cassidy, M.S.; Gary, E.N.; Love, J.; Franklin, M.C.; Height, J.J. Structures of human acetylcholinesterase in complex with pharmacologically important ligands. *J. Med. Chem.* **2012**, *55*, 10282–10286. [CrossRef] [PubMed]
15. Taylor, P.; Wong, L.; Radić, Z.; Tsigelny, I.; Brüggemann, R.; Hosea, N.A.; Berman, H.A. Analysis of cholinesterase inactivation and reactivation by systematic structural modification and enantiomeric selectivity. *Chem. Biol. Interact.* **1999**, *119–120*, 3–15. [CrossRef]
16. Hörnberg, A.; Tunemalm, A.K.; Ekström, F. Crystal structures of acetylcholinesterase in complex with organophosphorus compounds suggest that the acyl pocket modulates the aging reaction by precluding the formation of the trigonal bipyramidal transition state. *Biochemistry* **2007**, *46*, 4815–4825. [CrossRef] [PubMed]
17. Lewis, W.G.; Green, L.G.; Grynszpan, F.; Radić, Z.; Carlier, P.R.; Taylor, P.; Finn, M.G.; Sharpless, K.B. Click chemistry in situ: Acetylcholinesterase as a reaction vessel for the selective assembly of a femtomolar inhibitor from an array of building blocks. *Angew. Chem. Int. Ed. Engl.* **2002**, *41*, 1053–1057. [CrossRef]
18. Zha, X.; Lamba, D.; Zhang, L.; Lou, Y.; Xu, C.; Kang, D.; Chen, L.; Xu, Y.; Zhang, L.; De Simone, A.; et al. Novel Tacrine-Benzofuran Hybrids as Potent Multitarget-Directed Ligands for the Treatment of Alzheimer's Disease: Design, Synthesis, Biological Evaluation, and X-ray Crystallography. *J. Med. Chem.* **2016**, *59*, 114–131. [CrossRef] [PubMed]
19. Baykov, S.V.; Mikherdov, A.S.; Novikov, A.S.; Geyl, K.K.; Tarasenko, M.V.; Gureev, M.A.; Boyarskiy, V.P. π–π Noncovalent Interaction Involving 1,2,4- and 1,3,4-Oxadiazole Systems: The Combined Experimental, Theoretical, and Database Study. *Molecules* **2021**, *26*, 5672. [CrossRef] [PubMed]

20. Kryukova, M.A.; Sapegin, A.V.; Novikov, A.S.; Krasavin, M.; Ivanov, D.M. New Crystal Forms for Biologically Active Compounds. Part 1: Noncovalent Interactions in Adducts of Nevirapine with XB Donors. *Crystals* **2019**, *9*, 71. [CrossRef]
21. Rydberg, E.H.; Brumshtein, B.; Greenblatt, H.M.; Wong, D.M.; Shaya, D.; Williams, L.D.; Carlier, P.R.; Pang, Y.P.; Silman, I.; Sussman, J.L. Complexes of alkylene-linked tacrine dimers with Torpedo californica acetylcholinesterase: Binding of Bis5-tacrine produces a dramatic rearrangement in the active-site gorge. *J. Med. Chem.* **2006**, *49*, 5491–5500. [CrossRef] [PubMed]

MDPI
St. Alban-Anlage 66
4052 Basel
Switzerland
Tel. +41 61 683 77 34
Fax +41 61 302 89 18
www.mdpi.com

Crystals Editorial Office
E-mail: crystals@mdpi.com
www.mdpi.com/journal/crystals